NELSON KEY GEOGRAPHY

Foundations

5th Edition

DAVID WAUGH AND TONY BUSHELL

OXFORD
UNIVERSITY PRESS

OXFORD
UNIVERSITY PRESS

Great Clarendon Street, Oxford, OX2 6DP, United Kingdom

Oxford University Press is a department of the University of Oxford.
It furthers the University's objective of excellence in research,
scholarship, and education by publishing worldwide. Oxford is a
registered trade mark of Oxford University Press in the UK and in
certain other countries

Text © David Waugh, Tony Bushell 2014
Illustrations © Oxford University Press

The moral rights of the authors have been asserted

Key Geography Foundations published in 1991 and
Key Geography Foundations New Edition in 1996 by Stanley Thornes
(Publishers) Ltd.
Key Geography New Foundations (Third edition) published in 2001 and
New Key Geography Foundations (Fourth edition) in 2006 by Nelson
Thornes Ltd.
Fifth edition published in 2014

British Library Cataloguing in Publication Data
Data available

978-1-4085-2316-2

1 3 5 7 9 10 8 6 4 2

Printed in China

Acknowledgements

Page make-up: GreenGate Publishing Services, Tonbridge, Kent
Illustrations: Kathy Baxendale, Nick Hawken, Angela Knowles, Gordon
Lawson, GreenGate Publishing Services, Richard Morris, David Russell
illustration, Tim Smith, John Yorke.

The publishers would like to thank the following for permissions to
use their photographs:

Cover: Dudarev Mikhail/Shutterstock; Martin Harvey/Alamy; TACrafts/
iStockphoto **p4A:** Jane Rix/Shutterstock; **p5B:** Alberto Garcia/Corbis;
p5C: JonArnold Images/Photolibrary.com; **p5D:** Pavel Rahman/AP/
Press Association Images; **p14A i:** JL Images/Alamy; **p14A ii:** Mira/
Alamy; **p15A i:** Mitch Kezar/Getty Images; **p15A ii:** CandyBox Images/
Fotolia; **p16A:** Actionplus/Glyn Kirk; **p16B:** Jim Reed/Fdigital Vision
WW (NT); **p17C:** Corel 777 (NT); **p17D:** Brian Harris/Rex Features;
p18D i: Carles Zamorano Cabello/Alamy; **p18D ii:** Tony Bushell; **p18D
iii:** Ken Woodley; **p18D iv:** Adrian Bicker/Science Photo Library; **p21C:**
Tony Bushell; **p26B:** NERC Satellite Receiving Station, University of
Dundee; **p27E:** David Hughes/Shutterstock; **p28C:** NERC Satellite
Receiving Station, University of Dundee; **p33A:** Gorilla/Fotolia; **p33B:**
Brett Mulcahy/Fotolia; **p33C:** Maygutyak/Fotolia; **p33D:** Dan Race/
Fotolia; **p34A:** Digital Vision 15 (NT); **p34B:** Pavel Rahman/Empics/
AP; **p35C:** Colin Shepherd/Rex Features; **p35D:** John Giles/PA; **p36A
:**Tony Waltham; **p38A:** Photofusion/Rex Features; **p38B:** Jeff J Mitchell/
Getty Images News; **p40A i:** The Environment Agency; **p40A ii:** Steve
Speller/Alamy; **p40A iii:** Paul Rapson/Science Photo Library; **p40A iv:**
dszc/iStockphoto; **p40A v:** Jeff J Mitchell/Getty Images News; **p40A
vi:** Newsquest (Yorkshire & North East) Ltd; **p43 i:** Helen King/Corbis;
p43 ii: Construction photography; **p43 iii:** LWA; **p46A:** Lonely Planet
Images/Neil Setchfiled; **p47B:** Skyscan.co.uk/LAPL; **p47C:** Photofusion/
Bob Watkins; **p49C:** Airphotos; **p50A:** Skyscan.co.uk; **p50B:** Skyscan.
co.uk/CLI; **p50C:** Aerofilms/Alamy; **p53 i:** Heiko Butz/Fotolia; **p53 ii:**
godfer/Fotolia; **p53 iii:** micromonkey/Fotolia; **p53 iv:** iofoto/Fotolia;
p53 v: Dean Mitchell/iStockphoto, **p53 vi:** dundanim/Fotolia; **p53
vii:** Daniel Ernst/Fotolia; **p53 viii:** mangostock/Fotolia; **p57 i:** David
Waugh; **p57 ii:** Real Image/Alamy; **p57 iii:** Sean Aidan/Eye Ubiquitous;
p57iv: Rachel Dewis/iStockphoto; **p58A:** Tower Hamlets Local History
Library; **p58B:** Pictures Colour Library/ Alamy; **p60A:** Photofusion/
Rex Features; **p60B:** John Beacham/Eye Ubiquitous; **p62A:** Eric James/
Alamy; **p63C:** Terry Harris/Alamy; **p64A:** Guy Somerset/Alamy; **p64B
i:** Minerva Studio/Shutterstock; **p64B ii:** John Birdsall Photography/
Photofusion; **p67C:** Mike Robinson/Alamy; **p68B i:**Magalice/Fotolia;
p68B ii: Lsantilli/Fotolia; **p68B iii:** Andy Dean/Fotolia; **p68B iv:**
auremar/ Fotolia; **p69D:** Tony Bushell; **p72A:** NHPA/Photoshot; **p73B:**
John Warburton-Lee Photography/Alamy; **p73C:** Gavr iel Jecan/Corbis;
p73D: Images of Africa Photobank/Alamy; **p74A:** Dmitry Pichugin/
Fotolia; **p74B:** photoromano/Fotolia; **p74C:** Imagebroker/FLPA;
p76A: Barry Iverson/Alamy; **p76B:** Friedrich Stark/Alamy; **p76C:** age
fotostock/Robert Harding; **p79B:** Tony Bushell; **p79C:** Greg Evans
Photo Library; **p79D:** JDTP/Eye Ubiquitous; **p80A:** Robert Harding
Picture Library Ltd/Alamy; **p83C:** John Warburton-Lee Photography/
Alamy; **p84A:** Arterra Picture Library/Alamy; **p84B:** Crispin Hughes/
Panos Pictures; **p84C, p85D, p85E, p86A, p86B:** David Waugh;
p86C: Superstock/Age Fotostock/Tibor Bognár; **p87E:** Gary Cook/
Alamy; **p91D:** Andes Press Agency/Carlos Reyes-Manzo; **p103C i:** John
Heseltine/Corbis; **p103C ii, p103C iii:** Tony Bushell; **p103C iv:** Doug
Houghton/Alamy; **p104A:** Chris Schmidt/Getty Images; **p104B:** David
Gardner; **p105C:** John Giles/PA Archive/Press Association Images;
p106A: TTL/Photoshot; **p107C:** Nick Gregory/Apex; **p108 (top):**
Johnny Greig/iStockphoto; **p108B:** Digital Vision 15 (NT); **p110A:**
Andrew Brown, Ecoscene; **p111B:** Tony Bushell; **p112B:** Steven Vidler/
Eurasia Press/Corbis; **p113C:** London Aerial Photo Library; **p114B:** Air
Images; **p115D:** Blom Aerofilms Ltd/Science Photo Library; **p116B:**
Science Photo Library (Tom Van Sant/Geosphere Project); **p117C:** MDA
Information Systems/Science Photo Library.

Ordnance Survey maps (51D, inside back cover) reproduced by
permission of Ordnance Survey on behalf of HMSO.

Contents

1 What is geography?

Your passport to the world

What is this unit about?

This unit explains what is meant by geography and looks at some of the ways we study the subject. It also shows what you will learn from geography and how it can be valuable to you in later life.

In this unit you will learn about:

- differences between physical, human and environmental geography
- how to find places on a map
- how to use maps and photos to describe places
- how to understand and appreciate geography
- the value and use of geography.

Why is geography an important subject?

Geography is about people and places. It helps us understand our world and makes it a more interesting place in which to live. It helps us make sense of news events and what is going on around us. It also helps us understand ways of life that are different from our own and makes travel and meeting people more exciting.

Learning geography can also be of benefit to you in the future.

- It can give you an interest in people and places.
- It can introduce you to a variety of hobbies.
- It can give you job opportunities in a variety of interesting careers.

Put all this together and geography can be your passport to the world.

A Monument Valley, Utah, USA

B Mount Pinatubo eruption, Luzon, the Philippines

C Times Square, New York, USA

- Of the places shown in these photos, which
  - would you most like to visit
  - is the most different from where you live
  - is the most attractive
  - is the busiest
  - looks the most dangerous
  - would you like to know more about?

What are your reasons for your answers?

D Overcrowded train, Bangladesh

What is physical geography and ...

Physical geography is the study of the earth's natural features. It is about the land and the seas and the atmosphere around us.

The **atmosphere** is the air around the earth. Changes in temperature, rainfall and pressure give us our **weather** and **climate**. Climate changes between seasons and from year to year. Different parts of the world have different climates.

Landforms are natural features formed by rivers, the sea, ice and volcanoes. These are constantly changing as they are worn away in some places and built up in others.

The earth's surface is made up of many different kinds of rocks. Where these rocks break into small pieces they form soil. Plants grow in this soil and cover most of the earth's land surface as vegetation.

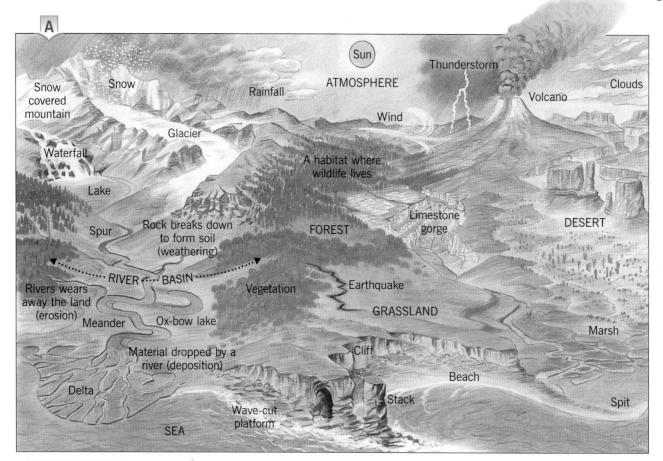

A

Snow covered mountain | Snow | Rainfall | ATMOSPHERE | Sun | Thunderstorm | Volcano | Clouds | Wind | Glacier | Waterfall | A habitat where wildlife lives | Lake | Spur | Rock breaks down to form soil (weathering) | FOREST | Limestone gorge | DESERT | Rivers wears away the land (erosion) | RIVER BASIN | Vegetation | Earthquake | GRASSLAND | Meander | Ox-bow lake | Marsh | Material dropped by a river (deposition) | Cliff | Beach | Delta | Stack | Spit | Wave-cut platform | SEA

Activities

1   a   Make a copy of table **C** below.
   b   List the features shown in drawing **A** in the three columns.

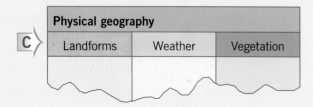

C

Physical geography		
Landforms	Weather	Vegetation

2   Copy and complete these sentences:
   a   Population geography is ...
   b   Settlement geography is ...
   c   Economic geography is ...

3   a   Make a larger copy of table **E**.
   b   Sort the statements from drawing **D** into the correct columns.
   c   Add at least two extra statements of your own to each column.

... what is human geography?

Human geography is the study of where and how people live.

Population geography looks at the distribution of people over the earth's surface. It looks at why people live in some places but not in others. It studies places where population is growing rapidly and looks at the problems that come with this growth. It also looks at why people migrate from one place to another and the effects that this movement has on an area.

Settlement geography is about where people live and how these places grow in size. It looks at how land is used in cities and how this can change over time.

Economic geography looks at how people earn a living. It is about industry, jobs and wealth.

Human geography also looks at **quality of life**. This is how happy or content people are with their lives and the environment where they live.

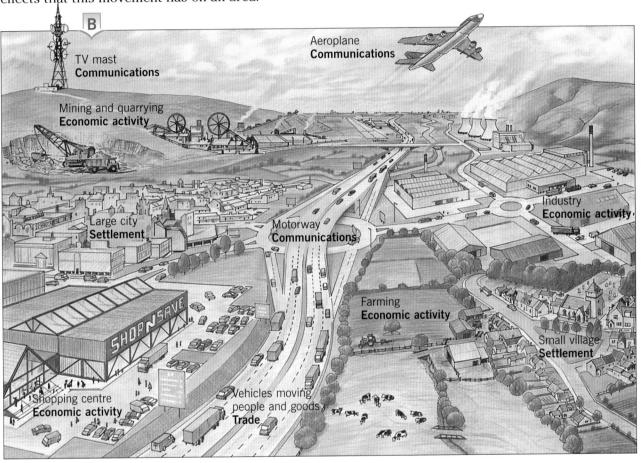

B
TV mast
Communications

Aeroplane
Communications

Mining and quarrying
Economic activity

Industry
Economic activity

Large city
Settlement

Motorway
Communications

Farming
Economic activity

Small village
Settlement

Shopping centre
Economic activity

Vehicles moving people and goods
Trade

D
- Looking for work
- Traffic jams in city centres
- Reasons for places to be crowded
- Problems in city centres
- How rich people are
- Land use in a city
- Immigrants moving into Britain
- Why few people live in desert areas
- Different beliefs and ways of life
- The location of shopping centres
- Why the UK is richer than Kenya
- Car making industry

E

Human geography		
Population	Settlement	Economic

4 a What is meant by the term 'quality of life'?
   b List six things that you think give people a good quality of life, and six things that give people a poor quality of life.

What is environmental geography?

A

Clean river

Fishing

Port

Sheltered bay protects ships from storms

Areas of scenic value attract tourists

Headlands for walks

Holiday resort with large hotels and amenities

Pier

The **environment** is the combination of the **physical** (natural) environment of climate, landforms, soils and vegetation, and the **human** environment which includes settlements and economic activities. It is the study of the surroundings in which people, plants and animals live.

The environment includes natural **resources** such as coal and iron ore, soils, forests and water. These are used to meet human needs. Some of these resources are **renewable**. This means that they can be used over and over again, such as rainfall. Others are **non-renewable** and can only be used once, such as coal. Sometimes people use these resources to their advantage. For example they use water for drinking purposes, iron ore in industry, and landforms such as islands or lakes for leisure. People often misuse these resources by using them up (minerals), by destroying them (soils, forests) or polluting them (rivers, seas and the air).

Different environments have different qualities and different uses. Each needs to be **protected** and carefully **managed**, like National Parks and the reserves of oil. Many environments have been damaged in the past. Those which have, such as mining areas, rivers and the older parts of some cities, need to be improved.

There is now a growing concern over the **quality of the environment** and how it may be **conserved** while at the same time being made as useful as possible.

B

Quarry in use

Old quarry hidden by trees

Trees chopped down

Smoke given off by factory

ENGWAL National Park

Soil washed away

Nature reserves, spits, dunes and marshes provide habitat for wildlife

Buildings on farmland

Dirty river with dead fish

Fumes given off by vehicles

Dirty beach with sewage outlet

Oil slick on sea

Untreated sewage

Coast

Activities

1 a Make a copy of table **C**.
   b List the features shown in drawing **A** in the two columns. The first two have been done for you.

2 In what ways has the area shown in drawing **B** been
   a polluted or destroyed
   b protected?

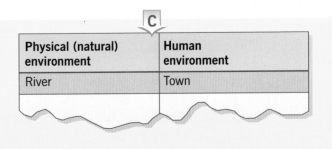

C

Physical (natural) environment	Human environment
River	Town

How can we find out where places are?

People often need to know where places are. They need to know this if, for example, they are going shopping or on holiday. Many people, like lorry drivers and ambulance drivers, need to know where places are to do their job. On television, radio and the internet we are always hearing about different places, on the news and in other programmes.

Geographers use maps to find out where places are and what they are like. An **atlas** is a book that has maps showing places all around the world, and it is easy to use. The most accurate way to show the whole world is on a **globe**. This is because a globe, being round, shows the actual shape of the earth.

To help us find places, imaginary lines called **latitude** and **longitude** are drawn onto the globe. These are shown in diagram **A**.

A

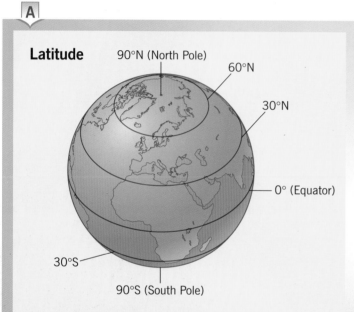

Lines of latitude are imaginary lines going around the earth from east to west. They are measured in degrees north or south of the **Equator**. Latitude 0° is called the Equator.

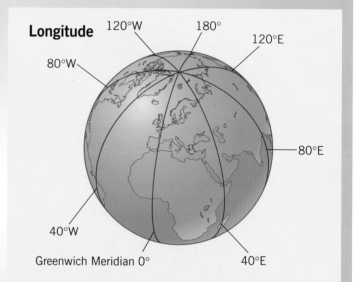

Lines of longitude are imaginary lines going from the North Pole to the South Pole. They are measured from the **Greenwich Meridian**, which passes through London. The Greenwich Meridian is 0°.

It is impossible to draw the earth accurately on a piece of paper. Parts of it will always be either the wrong size or the wrong shape. This is because the earth is round, and a piece of paper is flat. Imagine peeling the skin off an orange and trying to lay it out flat. It cannot be done, because the peel will split and some parts will be pushed out of shape.

One way of drawing the globe as a flat map is shown in map **B**. Some places have been stretched, and others squashed to make them fit.

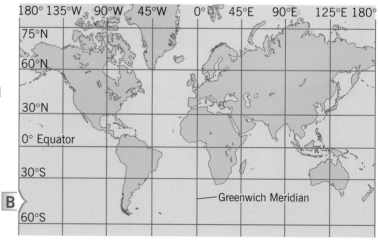

B

Using an atlas

The **contents** page at the front of the atlas shows on which page each map can be found. The **index** at the back of the book shows exactly where a particular place may be found. The index gives the latitude and longitude of that place to help you find it more easily. Diagram **C** shows you how to use the index of an atlas.

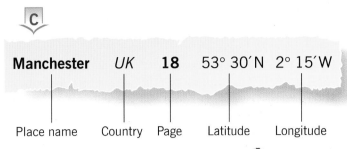

| Manchester | UK | 18 | 53° 30´N | 2° 15´W |

Place name   Country   Page   Latitude   Longitude

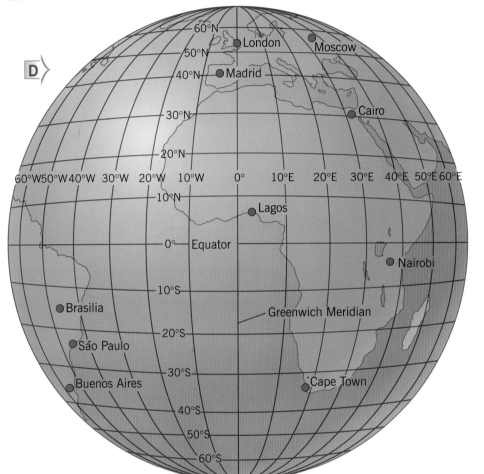

D

✔ **Remember**
- ✔ Lines of latitude go across the map.
- ✔ Lines of longitude go up and down the map.
- ✔ Latitude is always given first.
- ✔ Latitude and longitude are measured in degrees (°). Each degree is divided into 60 smaller parts called minutes (´).

Activities

1 Look at map **D** above. Name the city at each of the following:
   a 30°N 31'E   b 34°S 18'E
   c 40°N 4'W   d 24°S 47'W.

2 Use map **D** to give the latitude and longitude for each of these cities:
   a London   b Lagos
   c Moscow   d Buenos Aires
   e Nairobi   f Brasilia.

3 Find each of the cities below in the index of your atlas. For each one give the country, page number and latitude and longitude.

| New York | Tokyo | Sydney | Calcutta |

Summary

Maps are useful to people. They help us to find out where places are and what they are like. An atlas shows many places around the world. These places may easily be found using latitude and longitude.

How can we use graphs in geography?

Graphs are diagrams that show information in a clear and simple way. They can be used to describe a situation and show how one thing is related or linked to another.

Graphs are drawn using facts and figures which are called **data**. We can obtain data either by collecting information from fieldwork or by looking it up in a book. Information that we collect ourselves is called **primary data**. Information from other sources is called **secondary data**.

Graphs can either be drawn by hand or on a computer using a spreadsheet program.

✔ Remember

When you draw a graph, it should have:
- ✔ a title to say what it is showing
- ✔ labels along the bottom and up the side to explain what they are showing
- ✔ figures that are plotted very accurately.

Bar graphs A

Rainfall graph for London

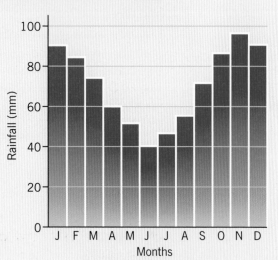

- A bar graph is made up of several bars or columns.
- The bars can be drawn either horizontally across the page or vertically up and down the page.
- Bar graphs are used to compare different things or quantities.
- The graph above compares the amount of rain in each month of a year. It shows which parts of the year are wettest.

Line graphs B

World population change

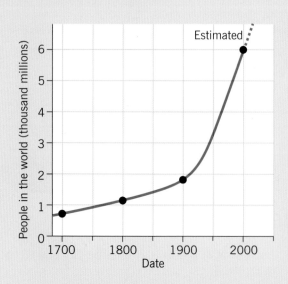

- Line graphs show information as a series of points that are joined up to form a line.
- Line graphs show changes or trends over a period of time. They can also help predict or forecast future changes.
- The graph above shows that population grew slowly between 1700 and 1800, then very rapidly between 1900 and 2000. The graph also suggests what might happen in the future.

Activities

1 Look at graph **A**.
  a Name the four months with least rainfall.
  b How much rainfall was there in November?

2 Look at graph **B**.
  a What was the population in 1700?
  b What does the graph suggest might happen to population in the future?

3 Look at graph **C**.
  a In which seasons are fewest holidays taken?
  b What percentage of holidays are taken in summer?

4 Look at graph **D**.
  a How much rainfall gave a river depth of 2 metres?
  b Which point, A, B or C, shows there to be little rainfall and a small amount of water in the river?

5 Which type of graph would be best for:
  a showing the change in the cost of petrol
  b comparing the size of UK cities
  c showing how the land use of an area is divided up?

Summary

Graphs are diagrams used to show data clearly. The four types of graph are the bar graph, line graph, pie graph and scatter graph.

Pie graphs

People taking holidays in the UK

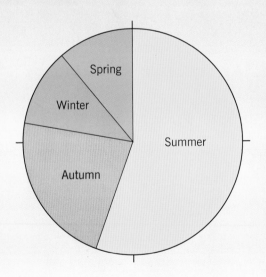

- A pie graph is drawn as a circle which is then divided into several pieces or sectors.
- The whole circle is always equal to 100%.
- Pie graphs show proportions and help us to see how something is divided up.
- The graph above shows that more than half of the people in the UK take their holidays in the summer months.

Scatter graphs

Rainfall and river depth

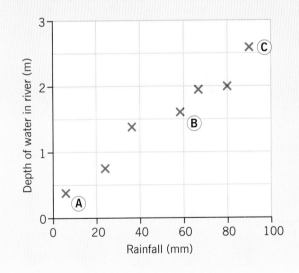

- A scatter graph has data plotted as a number of dots or crosses.
- Scatter graphs are used to see if information about two different things is related or linked.
- The graph above shows the link between rainfall and the amount of water in a river.
  As rainfall increases, so the amount of water in the river increases.

What is the value and use of geography?

The knowledge and skills that you learn in geography will help you in the future. They will give you an interest in people and places and help you understand what is going on in the world. They will also help you to make more sense of events in the news and enable you to develop your own views and opinions about both local and global issues.

As you will see, one of the best ways to learn about geography is to ask questions. Indeed you will notice that most pages in this book start with a key question and each chapter ends with an enquiry. Learning to ask questions and develop enquiry skills will help you find out things for yourself and make your own decisions.

But that's not all. The knowledge and skills that you learn in geography can also open the door to a variety of interesting and exciting careers. Jobs in travel, town planning, weather forecasting, mapping, journalism and the environment are just some of these.

Physical geography

Physical geography helps you:

- learn about the earth's landforms, climate and vegetation
- appreciate landscapes and scenery
- make sense of your surroundings
- be aware of what other places in the world are like
- know the causes of natural hazards
- know how to prepare for and cope with natural hazards.

Human geography

Human geography helps you:

- learn about and appreciate your surroundings
- learn what other countries and cities are like
- understand population growth and migration
- understand ways of life that are different from your own
- learn how and why countries are at different stages of development
- learn about some of the problems facing our world and how we might solve them.

Activities

1 Make a copy of the table below.

  a List the topics from drawing B in the correct columns.

  b Add at least two more topics to each column.

Some examples of topics in geography			
Physical	Human	Environmental	Skills

2 In what ways do you think geography will help you in your life? Try to give at least six.

3 Make a list of careers where the knowledge and skills that you learn in geography will be a help. Try to give at least ten.

- Planning walking routes
- Lake District scenery
- Migration and asylum seekers
- Local flooding in south-east England
- Pollution of rivers
- Using computers in ICT
- Indian Ocean earthquake and tsunami
- Traffic problems in your local area
- Global warming
- Damaging wildlife habitats

B

Summary

The knowledge and skills that you learn in geography can help you understand our world and will help you in future years.

Environmental geography

Environmental geography helps you:

- learn how we use the earth's natural resources
- understand what happens when we waste resources and damage the environment
- learn that we must live in a sustainable way
- learn how to recycle waste materials and reduce energy consumption
- develop a concern for the environment
- learn about protecting and conserving wildlife and scenery.

Skills in geography

Skills in geography help you:

- read maps and find your way about
- interpret and use graphs and statistics
- use maps and photos to find out what places are like
- learn how to carry out enquiries
- use questions and enquiry skills to find out things for yourself
- use computers (ICT) to find things out and present information.

2

How can the weather affect us?

What is this unit about?

This unit is about how weather and climate vary from time to time and from place to place. It shows how these variations are due to many different physical and human factors.

In this unit you will learn about:

- observing and recording the weather
- how local features affect temperature and wind
- what causes rain
- how weather and climate vary across Britain
- anticyclones and depressions
- how to forecast the weather.

A   A wet day at Wimbledon tennis championships, UK

B   Tree felled by a storm, East Anglia, UK

Why is this weather topic important?

Weather affects our lives in many ways. For example, it affects:

- the sort of activities we do
- the type of clothes we wear
- what we plan to do at the weekend
- where and when we go on holiday.

An understanding of the weather helps us make better use of weather forecasts and may even help us make predictions of our own. This can help us make the best use of weather conditions and to avoid the problems that unexpected changes in the weather can bring. This unit will help you to do that.

- For each photo, how would
  - the weather affect you
  - you feel about being there
  - you prepare for that weather?
- What problems may these types of weather cause?

C Canary Islands

D Traffic chaos in a UK city

How might you observe and record the weather?

Weather can be described as the condition of the air around us over a short period of time. It is about being hot or cold, wet or dry, windy or calm, cloudy or sunny.

Meteorology is a study of the weather. One of the important tasks of meteorologists is to measure and record all the features of the weather every day. Many expensive and complicated instruments are needed to record weather accurately but you can get a good picture of what conditions are like by **observing** (looking around) and using simple equipment.

A

Temperature
This is a measure of how hot or cold it is. You can do this by looking at the clothes that people are wearing. Thermometers are used to measure temperature accurately.

| Very cold | Cold | Mild | Warm | Hot |

B

Precipitation
Water in the air falls to the ground in one of several forms. Four of these are rain, snow, sleet and hail.

C

Wind speed
This tells us how strong the wind is. We can get a good idea of this by looking at smoke and the trees. The **Beaufort scale** is used to measure wind strength.

| 0 Calm | 2 Light breeze | 4 Moderate breeze | 6 Strong breeze | 8 Fresh gale |
| Smoke rises vertically | Wind felt on face, leaves rustle | Dust and paper lifted, small branches move | Large branches in motion | Twigs break off trees |

D

Cloud type
Cloud comes in many shapes, sizes and heights. Cumulonimbus, cumulus, stratus and cirrus are the most common types.

Cumulonimbus

Cumulus

Stratus

Cirrus

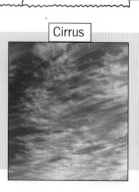

Wind direction
This is the direction **from** which the wind blows. It is shown by a wind vane.

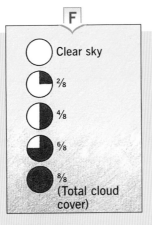

Cloud cover
This is the amount of the sky covered by cloud. It is measured in eighths.

Clear sky
²⁄₈
⁴⁄₈
⁶⁄₈
⁸⁄₈ (Total cloud cover)

Visibility
This is the distance that can be seen. It is measured in metres.

General weather
This describes the weather in words. Words like rain, snow, showers, fog, mist, thunder, cloudy, fair or sunny are used. Light or heavy can be added to precipitation.

Activities

1 What is weather?

2 a Make a copy of diagram **I** on the right.
   b Write the name of the weather feature next to each sketch.

3 Describe how each of the following is measured:

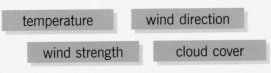

temperature   wind direction

wind strength   cloud cover

4 Make sketches of the four cloud types in **D**. Under each sketch write a cloud description from the following list.
   • Low grey shapeless cloud that forms in layers.
   • High clouds that are wispy, light and featherlike.
   • Dome shaped clouds with dark flat bases.
   • Huge towering clouds that often give showers.

Weather features to be observed and recorded

5 Look at table **J** on the right which shows what the weather was like on a summer day in Wales.
   a Copy out the table headings.
   b Make your own recording of today's weather. Use the information on these two pages to help you.

6 a Keep a record of the weather for a week. Do this at the same time each day.
   b Record your readings in a table.

7 See if you can spot any link between the wind direction and other features of the weather.

J

Day	Temperature	Precipitation	Wind speed	Wind direction	Cloud amount	Cloud type	Weather
Sunday 15 July	Warm	Rain showers	Force 2	Westerly	4/8	Cumulus	Mainly sunny with some rain
Monday 16 July							

Summary

Weather is the day to day condition of the atmosphere. A simple record of the weather may be made by careful observation of what is going on around us.

How can local features affect temperature and wind?

On a fine summer's day, are some of the classrooms in your school hotter than others? When the sun shines or a cold wind blows, is one side of your classroom warmer or colder than the other? On a hot sunny day can you notice a difference in temperature between a dark, tarmac playground and a grassy area like the school field? Are there some sheltered places around your school where you can get out of the wind?

Look at cartoon **A** which shows how different the conditions can be on two sides of a hedge.

Each particular place or site tends to develop its own special climate conditions. When the climate in a small area is different from the general surroundings it is called a **microclimate**. Some of the causes of microclimates are given in **B** below.

A

Physical features

Trees provide shade and shelter and are usually cooler than surrounding areas. Water areas such as lakes and seas have a cooling effect and may also produce light winds. Hilltops are usually cool and windy.

Buildings

Buildings give off heat that has been stored from the sun during the day or which leaks from their heating systems. Temperatures near buildings may be 2°C or 3°C higher.

Buildings break up the wind and can reduce wind speeds by up to a third. Sometimes the wind can increase speed as it rushes around buildings.

B

Shelter

Trees, hedges, walls, buildings and even hills can provide shelter from the wind. Wind speed may be reduced and its direction changed. Places sheltered from cold winds will be warmer.

Surface

The colour of the ground surface affects warming. Dark surfaces such as tarmac and soil will become warmer than light surfaces such as grass.

Aspect

The direction in which a place is facing is called its aspect. Places facing the sun will be warmer than those in shadow.

In Britain the sun rises in the east and moves through the south before it sets in the west. South-facing places get most of the sun and are usually the warmest.

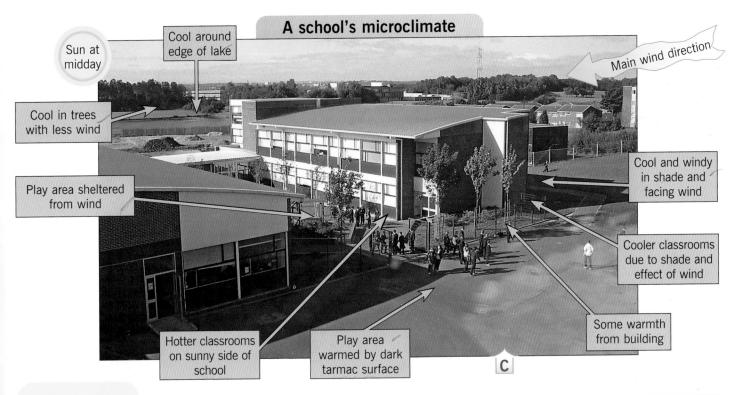

A school's microclimate

Sun at midday

Cool around edge of lake

Main wind direction

Cool in trees with less wind

Play area sheltered from wind

Cool and windy in shade and facing wind

Cooler classrooms due to shade and effect of wind

Hotter classrooms on sunny side of school

Play area warmed by dark tarmac surface

Some warmth from building

C

Activities

1 Describe a place at your school which is
   a often sunny
   b usually in the shade
   c sheltered on a windy day.

2 Copy and complete diagram **D** by filling in the clouds with the following words or statements:
   • Climate conditions of a small area
   • Physical features
   • Dark surfaces warm up most
   • Reduces the effect of wind
   • Buildings
   • Aspect

3 From photo **C** give **eight** features of the school's microclimate. List your answers under the headings:
   • Aspect   • Shelter   • Others

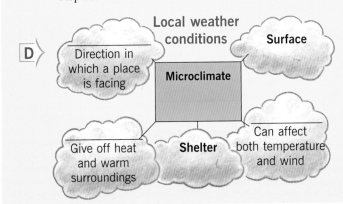

D

Local weather conditions

Direction in which a place is facing

Surface

Microclimate

Give off heat and warm surroundings

Shelter

Can affect both temperature and wind

4 **Microclimate Enquiry**

• **Aim** – to find out what effect aspect has on temperature

• **Equipment** – thermometer

• **Method**
   a Take several temperature readings on the north and south facing sides of the school. Make a note of the weather each time (e.g. sunny, cloudy, windy).
   b Make a copy of the table below and display your results.
   c Describe your findings.
   d Suggest reasons for your findings.

Time	North facing	South facing	Weather conditions
Average			

Summary

Site conditions such as aspect, shelter, physical features and other factors can influence temperature, local wind speed and direction.

What is Britain's weather?

Weather is what happens in the atmosphere day by day but **climate** is different. It is the weather taken on average over many years. Climate is about warm dry summers, cool wet winters or, as at the North and South Poles, being cold all year. In Britain the weather is always a popular topic of conversation probably because it is always changing or it's never quite what we want it to be. Changes also occur in the climate. It can change from time to time (seasonal) or it may be different from place to place.

Temperature

The average monthly temperatures for summer and winter are shown on maps A and B. If you look closely you should see three main differences. These differences are explained below.

1 As expected, temperatures are higher in summer than in winter.

2 Temperatures at any one time are not the same all over Britain.

3 The pattern of temperature is different in the two seasons.

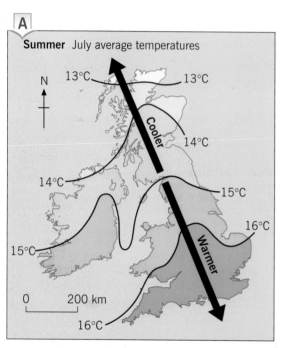

A **Summer** July average temperatures

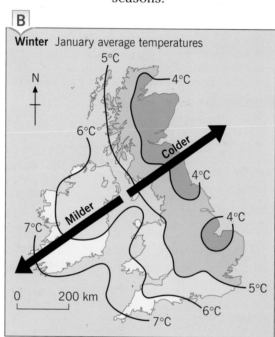

B **Winter** January average temperatures

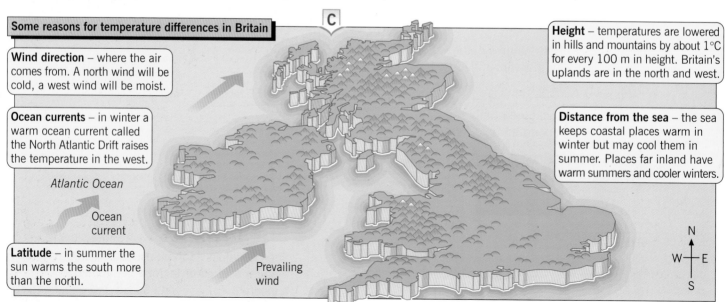

C

Some reasons for temperature differences in Britain

Wind direction – where the air comes from. A north wind will be cold, a west wind will be moist.

Ocean currents – in winter a warm ocean current called the North Atlantic Drift raises the temperature in the west.

Atlantic Ocean

Ocean current

Latitude – in summer the sun warms the south more than the north.

Prevailing wind

Height – temperatures are lowered in hills and mountains by about 1°C for every 100 m in height. Britain's uplands are in the north and west.

Distance from the sea – the sea keeps coastal places warm in winter but may cool them in summer. Places far inland have warm summers and cooler winters.

Rainfall

In Britain we can expect rain at any time of the year. Although winter is wetter than the summer, seasonal differences in rainfall are very small.

As map **D** shows, however, the amount of rainfall varies considerably from place to place and the greatest differences are between the east and the west.

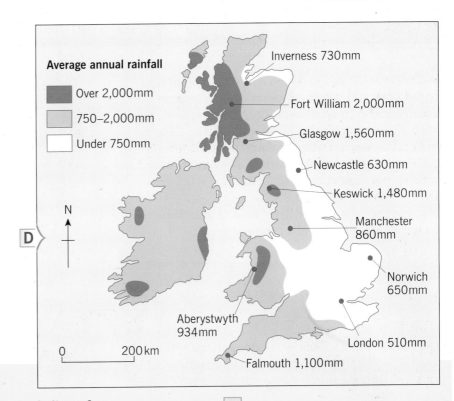

D

Activities

1. What is the difference between weather and climate?

2. a. Write out and complete the following sentence to describe summer temperatures in Britain.

   > Summers in Britain are _____ than winter. The warmest weather is in the _____ and temperatures get lower (decrease) towards the _____ .

   b. Write a similar sentence to describe winter temperatures.

3. Why are there temperature differences in Britain? Think of **three** reasons and write them in your workbook.

4. a. List the **three** wettest and the **three** driest towns from map **D**. Give your answers in order with the wettest first.

   b. With the help of a simple diagram, describe the difference in rainfall from east to west. Give actual figures in your answer.

5. a. Make a large copy of map **E**.

   b. Match the following climate descriptions to Ⓐ, Ⓑ, Ⓒ, and Ⓓ and write them on your map. Ⓐ, has been done on the map to help you.
   - Warm summers, cold winters, dry
   - Mild summers, mild winters, wet
   - Warm summers, mild winters, quite wet
   - Mild summers, cold winters, dry

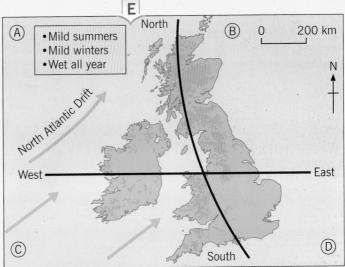

E

c. Suggest reasons for the climate of area Ⓐ.

d. Mark where you live on your copy of map **E**.

e. Describe the climate there and suggest reasons for it.

Summary

Britain's climate varies from place to place and from season to season. Heating from the sun, ocean currents, and the height of the land are some of the reasons for these variations.

23

How does it rain?

The Atacama Desert in South America has had no rain for over 400 years yet parts of the Amazon rainforest, also in South America, have rain on more than 330 days each year. Seathwaite in the Lake District, the wettest place in England, has on average 3,340 mm of rain per year, whilst Newcastle, only 130 km away, may expect just 630 mm.

What are the reasons for this, what causes rain, and why are some places wetter than others?

Clouds are made up of extremely tiny drops of moisture called **cloud droplets**. They are only visible because there are billions of them crowded together in a cloud.

Clouds form when moist air rises, cools and changes into cloud droplets. This is **condensation**. A cloud gives rain after these tiny cloud droplets grow thousands of times larger into raindrops which then fall to the ground.

Look at diagram **A**. It shows how rain is formed. The process is always the same: air rises, cools, condenses and precipitates.

Air can be forced to rise in three different ways.

This gives the three main types of rainfall: **relief**, **convectional** and **frontal**. These are shown in diagrams **B**, **C** and **D**.

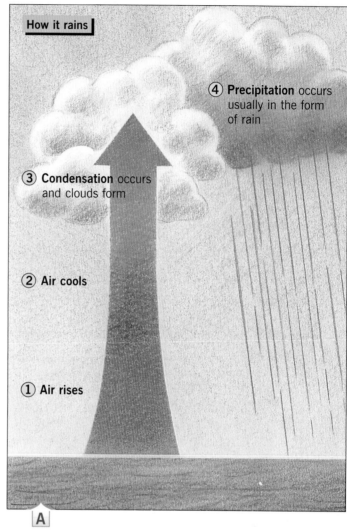

How it rains

④ **Precipitation** occurs usually in the form of rain

③ **Condensation** occurs and clouds form

② **Air cools**

① **Air rises**

A

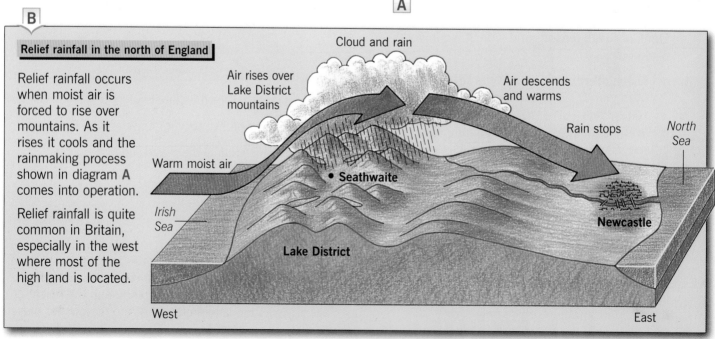

B

Relief rainfall in the north of England

Relief rainfall occurs when moist air is forced to rise over mountains. As it rises it cools and the rainmaking process shown in diagram **A** comes into operation.

Relief rainfall is quite common in Britain, especially in the west where most of the high land is located.

Cloud and rain

Air rises over Lake District mountains

Air descends and warms

Rain stops

North Sea

Warm moist air

• Seathwaite

Irish Sea

Lake District

Newcastle

West

East

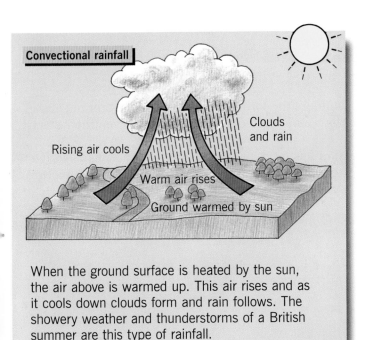

Convectional rainfall

Clouds and rain

Rising air cools

Warm air rises

Ground warmed by sun

When the ground surface is heated by the sun, the air above is warmed up. This air rises and as it cools down clouds form and rain follows. The showery weather and thunderstorms of a British summer are this type of rainfall.

C

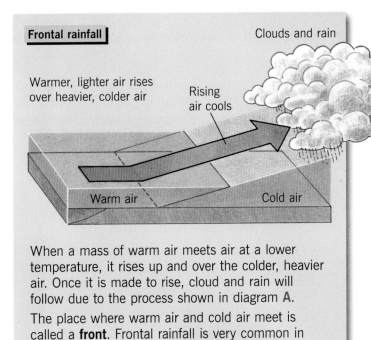

Frontal rainfall

Clouds and rain

Warmer, lighter air rises over heavier, colder air

Rising air cools

Warm air

Cold air

When a mass of warm air meets air at a lower temperature, it rises up and over the colder, heavier air. Once it is made to rise, cloud and rain will follow due to the process shown in diagram **A**.

The place where warm air and cold air meet is called a **front**. Frontal rainfall is very common in Britain throughout the year and especially in winter.

D

Activities

E

1   Match the beginnings of the labels in **E** to their correct endings.

2   With the help of a labelled diagram, describe how it rains.

3   a   Make larger copies of the three diagrams in **F**.

    b   For each diagram explain how it rains by adding labels at points ①, ②, ③ and ④.

    c   Add colour to make your diagrams clearer.

    d   Underneath each of your diagrams give a brief reason for the air rising.

    e   Give each diagram a title.

Clouds are | rain, snow and other forms of moisture in the sky.

Precipitation is | when water vapour changes to water.

Condensation happens | made up of tiny drops of moisture called cloud droplets.

4   Explain why Seathwaite is wetter than Newcastle. Use diagram **B** to help you.

F

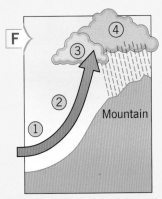

Mountain

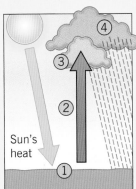

Sun's heat

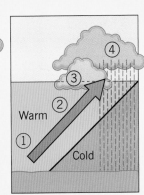

Warm

Cold

Summary

Rain is caused by moist air rising and cooling. The three types of rainfall produced in this way are relief, convectional and frontal.

Forecasting the weather – anticyclones

Weather has an important effect on our lives. Every day in the newspapers and every evening after the television news there is a **weather forecast**. Forecasts can tell us in advance what the weather will be. For many of us they are of passing interest but for some people such as farmers, fishermen, aircraft pilots and builders the forecasts are very important because the weather affects their work and even their safety.

Map **A** is a typical newspaper weather map. Notice how easy it is to read the weather using the picture symbols.

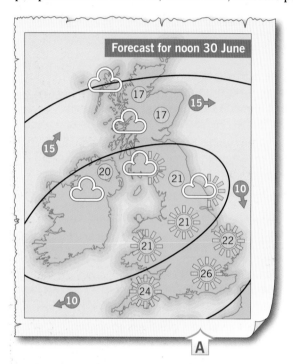

Forecast for noon 30 June

A

B Satellite photo of an anticyclone

How do weather forecasters know what the weather will be like tomorrow? How can they tell if it will be wet or dry, or hot or cold?

Forecasting is very complicated and lots of information and advanced computers are needed to make good forecasts. In recent times, satellites have become particularly useful because they can see weather systems many kilometres away.

Photo **B** was taken from a satellite. It shows Britain with very little cloud overhead and clearly enjoying a fine sunny day. Photos like these are taken every few hours and by looking back over several of them the movements of the weather systems can be worked out, and forecasts made.

The weather system in photo **B** is an **anticyclone**. It occurs because of changes in the air pressure. The weight of air pressing down on us from above is called pressure. This pressure varies from place to place and results in the development of pressure systems. Areas with above average pressure (high pressure) are called anticyclones and usually give good weather. Areas with less than average pressure (low pressure) are called depressions and usually give poor weather.

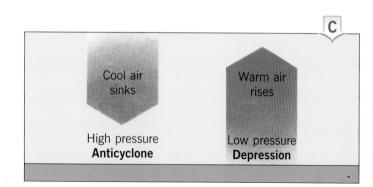

C

Cool air sinks

High pressure
Anticyclone

Warm air rises

Low pressure
Depression

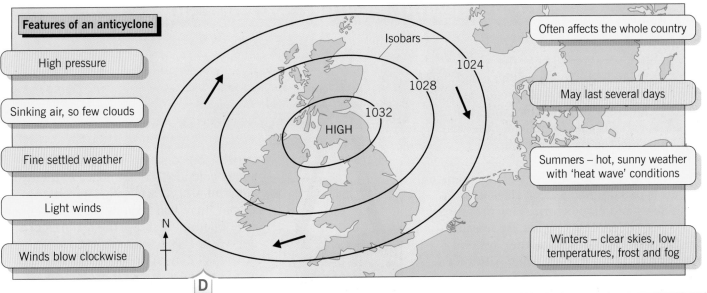

Features of an anticyclone

High pressure

Sinking air, so few clouds

Fine settled weather

Light winds

Winds blow clockwise

Isobars

1024

1028

1032

HIGH

N

Often affects the whole country

May last several days

Summers – hot, sunny weather with 'heat wave' conditions

Winters – clear skies, low temperatures, frost and fog

D

Activities

1 From map **A**, give the weather that is forecast for the place where you live.

2 a When do you think it would be useful for you to know the next day's weather?

 b Make a list of people who need the weather forecast. For each person explain why they need to know about the weather.

3 How do satellites help in forecasting weather conditions?

4 a Make a sketch of an anticyclone like the one in diagram **D** above.

 b Next to your sketch, write out the paragraph below and fill in the blank spaces with the following words:

> • LONG • LARGE • HIGH • COOL
>
> Anticyclones are areas of _____ pressure which form when _____ air sinks. They usually cover _____ areas and give _____ periods of fine settled weather.

5 Copy and fill in table **F** to show the weather features of an anticyclone.

E Weather in a winter anticyclone

6 Study map **A** on page 26 giving the newspaper weather forecast. Write a weather forecast to be read out on the radio for the same day. Your forecast should be about 100 to 150 words in length.

F

	Summer	Winter
Temperatures		
Cloud cover		
Wind speed		
Wind direction		
Rain		
Other features		

Summary

Knowing what the weather will be like can be useful to us. Anticyclones can bring good weather and may be forecast with the help of satellites.

Forecasting the weather – depressions

All too often we seem to hear the weather forecast begin with 'Today will be cloudy, and rain already in the west will spread eastwards to cover all areas by late afternoon.' The reason for this is that for much of the year Britain is affected by low pressure.

As diagram **A** shows, at times of low pressure the air is usually rising. As it rises it cools, condenses and clouds form. Low pressure areas are called **depressions**. Depressions are the most important weather systems affecting Britain and they bring with them clouds and rain.

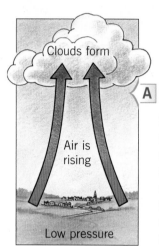

Depressions develop where warm air meets cold air. The boundary of the two different air types is called a **front**. Along a front there will be cloud and usually rain. Diagram **B** shows the features of a depression. The **isobars** are lines that join up areas of equal pressure and they help us to see the shape of the depression.

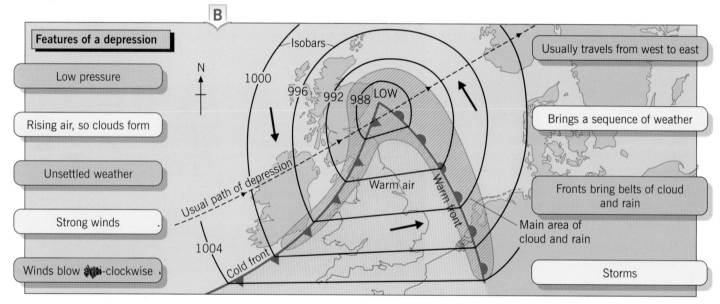

Features of a depression

Low pressure

Rising air, so clouds form

Unsettled weather

Strong winds

Winds blow anti-clockwise

Usually travels from west to east

Brings a sequence of weather

Fronts bring belts of cloud and rain

Storms

Main area of cloud and rain

Usual path of depression

Warm air

Warm front

Cold front

Isobars

N

1000   996   992   988   LOW   1004

C Satellite photo of a depression

Depressions are huge areas of low pressure measuring many hundreds of kilometres across. They show up very clearly on satellite photographs as great swirls of cloud that look like gigantic catherine wheel fireworks. The fronts are easily recognised as areas of thick white cloud arranged in an upside down 'V' shape. The centre of the depression is normally just above or a little behind the point of the 'V'.

Look at photo **C** which shows a depression approaching Britain. Can you work out which areas are the fronts and where the centre of the depression might be? With help from diagram **B** can you work out which is the area of warm air? What sort of weather does that area seem to have?

Depressions usually form over the Atlantic Ocean and move across Britain from west to east. With help from satellite photographs, weather forecasters can work out the direction they are travelling and how fast they are moving. From this information they can produce quite accurate weather forecasts. Diagram **D** shows how the weather changes as a depression passes over Britain. Notice the changes in the weather that occur in the area where you live.

D A depression passing over Britain

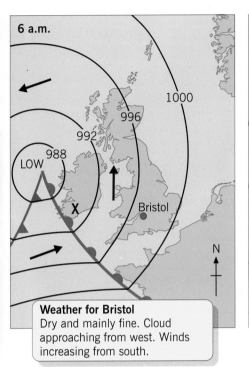

6 a.m.

Weather for Bristol
Dry and mainly fine. Cloud approaching from west. Winds increasing from south.

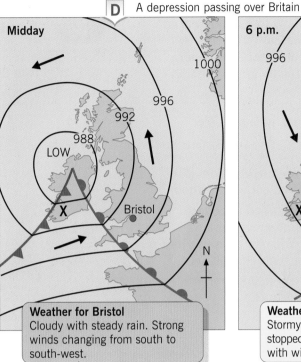

Midday

Weather for Bristol
Cloudy with steady rain. Strong winds changing from south to south-west.

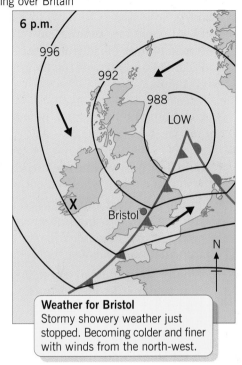

6 p.m.

Weather for Bristol
Stormy showery weather just stopped. Becoming colder and finer with winds from the north-west.

Activities

1 The words below have been jumbled up. Unscramble the words and fill in the blank spaces in the following paragraph.

NIRA	SATE	DULOC	OWL	TEWS	SERIS

Depressions are areas of _____ pressure which form when air _____ . They usually move across Britain from _____ to _____ and bring most of our _____ and _____ .

2 With the help of a labelled diagram, explain why depressions bring cloud and rain.

3 Make a labelled sketch of a depression like the one shown in diagram **B**. Underneath your sketch make a copy of the table below.

Complete the table to show the main features of a depression.

General features	Weather

4 From diagram **D**:
   a Describe the weather at place **X** for 6 a.m., 12 midday and 6 p.m.
   b Explain why the weather has changed.
   c At what time will the warm front be over the place where you live?
   d Describe the weather you may get at that time.

5 a Trace the outline of Britain from photo **C**.
   b Mark and label the following:
      • warm front
      • cold front
      • warm air
      • centre of the depression.
   c Shade the area of cloud and rain along the fronts.
   d Describe the weather over Britain.

Summary

Depressions are the most common weather system affecting Britain. They are low pressure areas and bring stormy winds, cloud and rain.

This enquiry is concerned with the weather and climate. Pages 22 and 23 of this book may be helpful to you as you work through the enquiry. Your task is to reply to a letter sent to you by a company in America. To do this you will need to look closely at Britain's weather and climate and answer the enquiry question given below.

There should be three main parts to your enquiry.

- The first part will be an introduction. Here you can explain what the enquiry is about.
- In the next part you will need to collect and present information about Britain's weather and climate. You can then use that information to answer the question set.
- Finally you will need a conclusion. Here you could write a letter to explain your findings.

What are the differences in weather and climate across Britain?

1 **Introduction – what is the enquiry about?**

You will need to use maps and writing here.

Star diagrams or lists might also help.

a First look carefully at the enquiry question above and say what you are going to try to find out.

- Give the meanings of **weather** and **climate**. Pages 18, 22 and the Glossary will help you.
- Describe how Britain's weather and climate can be roughly divided into four regions.
- Explain where these regions are, and then describe the different conditions in them.

b Describe briefly what the letter has asked you to do. Show on a map where the four places are located. List the particular features of weather and climate that you will look at.

World Wide Leisure Corporation

174 Aspen Boulevard, Denver Colorado 96541, USA

Tel/Fax (303)569-3309

Dear Sir/Madam

I am the Personnel Manager for a large American company. We are planning to open four offices in Britain. These offices will be at Oban, Aviemore, Plymouth, and Cambridge. Each manager will bring the family with them, and they are likely to stay for three years.

Like many Americans, each family is keen on leisure and doing things out of doors. Of course these activities in turn depend upon the weather and climate. Each family has different interests – as will be listed later in this letter. We are therefore allowing each of them to choose the office in the region where the weather and climate best suits their interests.

To help them do this we would appreciate your help. Please could you give us the weather and climate for the four places and suggest which you think is best suited to each of the families.

Yours sincerely

John F. Gates
Personnel Manager

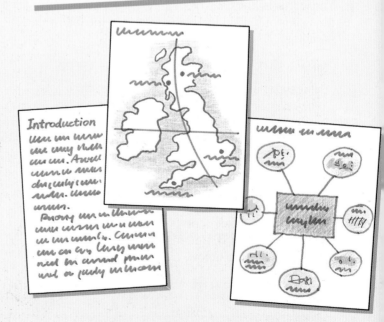

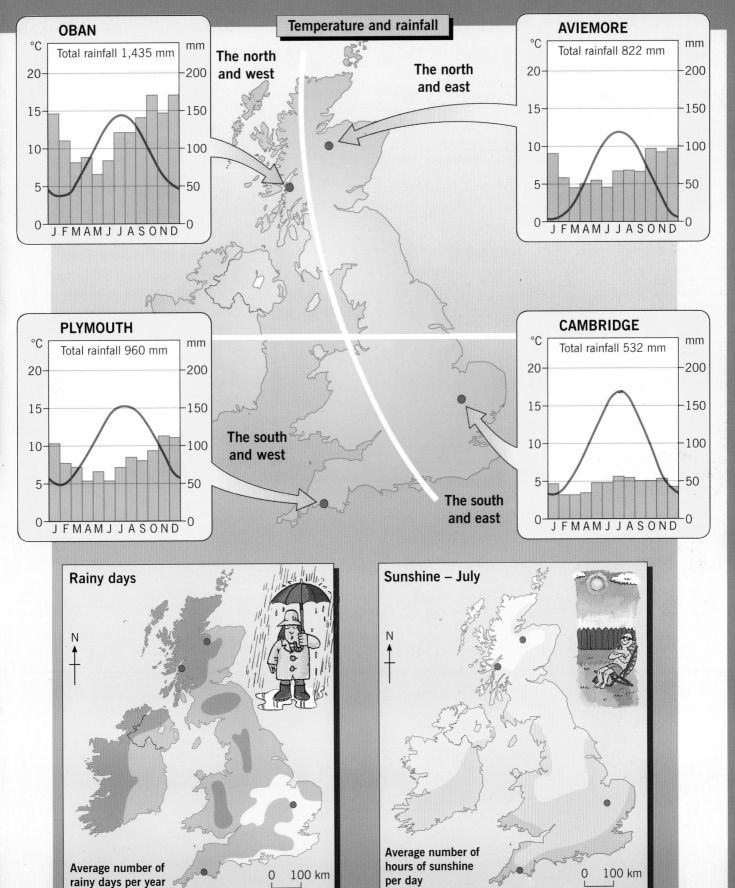

Temperature and rainfall

OBAN
Total rainfall 1,435 mm

The north and west

The north and east

AVIEMORE
Total rainfall 822 mm

PLYMOUTH
Total rainfall 960 mm

The south and west

The south and east

CAMBRIDGE
Total rainfall 532 mm

Rainy days

N

Average number of
rainy days per year

0   100 km

225 and over   175 – 224   Below 175

Sunshine – July

N

Average number of
hours of sunshine
per day

0   100 km

Over 7   6 – 7   5 – 6   Below 5

The weather enquiry

2 What is Britain's weather and climate like?

a Make a large copy of the table below of Britain's weather.

b Complete the table using information from this page and from page 31.

Britain's weather	Oban (north and west)	Aviemore (north and east)	Plymouth (south and west)	Cambridge (south and east)
January temperature (°C)				
July temperature (°C)				
January rainfall (mm)				
July rainfall (mm)				
Total rainfall (mm per year)				
Rainy days (number per year)				
July sunshine (hours per day)				
Snow lying (days per year)				
Average wind strength (description and km/h)				

3 Where is the best weather?

Each family made a list of the weather and climate that they would like to have for their stay in Britain. This information is given on page 33. You can now find out which places are most suited to each family.

a Make a copy of the table for the Jackson family.

b For each place in turn put:
   ✓ a tick if the weather is suitable
   ✗ a cross if it is unsuitable
   ? a question mark if it is not perfect but not too bad.

Your completed table showing Britain's weather will give you all the answers for this.

c Add up the ticks to find which place is the most suitable. The one with the most ticks would be the best.

d Now repeat the parts a, b and c for each of the other three families.

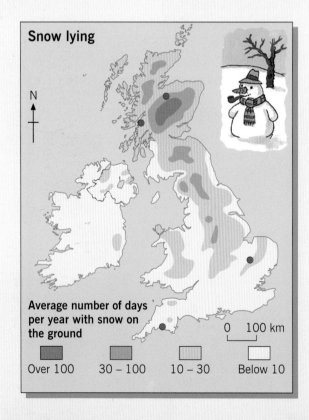

Snow lying

N

Average number of days per year with snow on the ground

0   100 km

| Over 100 | 30 – 100 | 10 – 30 | Below 10 |

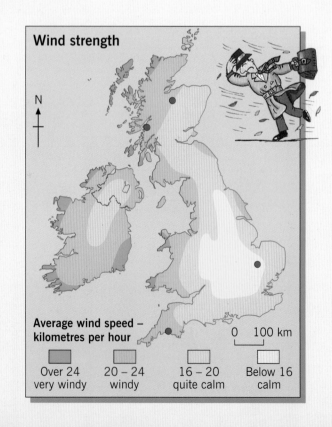

Wind strength

N

Average wind speed – kilometres per hour

0   100 km

| Over 24 very windy | 20 – 24 windy | 16 – 20 quite calm | Below 16 calm |

We are a cycling family so we don't like rain or wind. We prefer warm summers and cold winters.

We prefer it not to be cold or too snowy. We like to go fishing so rainy days can be good for us.

The Jackson family	Oban (north and west)	Aviemore (north and east)	Plymouth (south and west)	Cambridge (south and east)
Cold winters (Jan. temp. below 3°C)				
Warm summers (July temp. 15–20°C)				
Dry (less than 175 rainy days)				
Quite sunny in summer (6–7 hrs per day)				
Very little wind (below 16 km/h)				
TOTAL				

The Houston family	Oban (north and west)	Aviemore (north and east)	Plymouth (south and west)	Cambridge (south and east)
Mild winters (Jan.temp. 3–7°C)				
Mild summers (July temp. 10–14°C)				
Many rainy days (over 225 per year)				
A little snow (10–13 days per year)				
Windy (20–24 km/h)				
TOTAL				

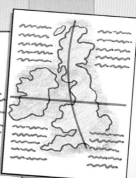

We are keen walkers and skiers. Our favourite days are in winter when it is cold and snowy.

We enjoy barbecues and relaxing in the sun. We like warm sunny summers. Rain doesn't bother us but we really don't like the cold.

The Grant family	Oban (north and west)	Aviemore (north and east)	Plymouth (south and west)	Cambridge (south and east)
Cold winters (Jan. temp. below 3°C)				
Mild summers (July temp. 10–14°C)				
Quite dry (total rain 600–900 mm)				
Cloudy summers (under 5 hrs per day)				
Lots of snow (over 30 days per year)				
TOTAL				

The Stolberg family	Oban (north and west)	Aviemore (north and east)	Plymouth (south and west)	Cambridge (south and east)
Mild winters (Jan. temp. 3–7°C)				
Warm summers (July temp. 15–20°C)				
Quite wet (total rain 900–1,200 mm)				
Lots of summer sunshine (over 7 hrs per day)				
Windy (20–24 km/h)				
TOTAL				

4 Conclusion

Now you should look carefully at your work and answer the enquiry question. You could do this by replying to the letter from the World Wide Leisure Corporation. This could include writing and perhaps a labelled map.

a Describe the weather and climate in each of the four regions of Britain. Your completed tables from this page will help you.

b Say which place is best suited to each of the four families. Give reasons for your answer.

3

Why is flooding a problem?

What is this unit about?

This unit is about the causes and effects of river flooding. It shows how flooding can affect people in different ways and suggests how the risk of future flooding may be reduced.

In this unit you will learn about:

- what happens to rain when it reaches the ground
- the causes of flooding
- how individuals and communities respond to the problem of flooding
- how the risk of flooding in the UK can be reduced.

A | Extensive flooding in Bangladesh

B | A flooded slum, Mirpur, Bangladesh

Why is this flooding topic important?

Flooding is an increasing problem across the world. More than 7 million people in the UK are now at risk from flooding every year. Even if flooding has not affected you yet, it could easily do so some time in the future. For these reasons, we need to understand the causes and effects of flooding so that we can try to manage the problems they create.

This unit can also help you to:

* be aware of the effects of flooding
* find out if flooding is a problem where you live
* prepare for a flooding situation
* know what to do during and after a flood
* be able to help other people affected by flooding.

* Look at the people in the photos.
  – What problems do they face?
  – What help do they need?
  – Who needs the most help?
  – How could they prepare for the situation?
  – What could you do to help them?

C Cleaning up after flooding, Boscastle, UK

D Rescue from floods, Carlisle, UK

What causes a river to flood?

All of the water that flows down a river comes from rain or melting snow. Sometimes after heavy rain or a rapid snow melt, there may be too much water for the river to hold. The river will then overflow its banks and spread out across the land on either side of its channel. This is called a **river flood**.

Usually when it rains, most water simply soaks into the ground and there is little chance of a flood. If, however, the water is unable to soak into the ground, it will stay on the surface and flow quickly downhill and into the river. This is when floods are most common.

Some rivers are more at risk from flooding than others. Put simply, heavy rain and anything which stops that rain from soaking into the ground will increase the chances of flooding. Some of the factors that increase the risk of flooding are shown below.

A

The rapid melting of snow can cause flooding

B

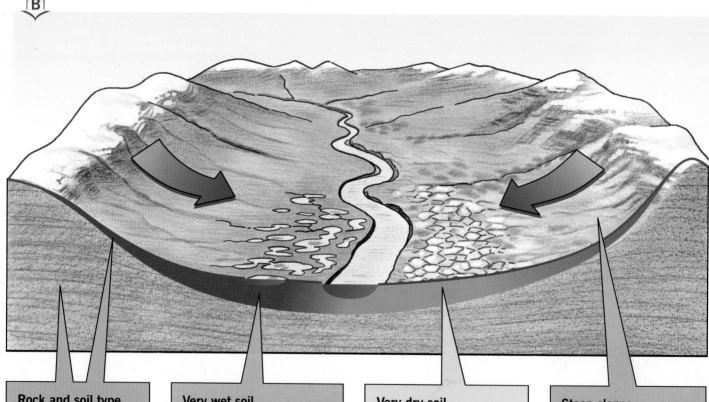

Rock and soil type
Impermeable rocks and soils do not allow rain to soak through them. Any rain that falls will stay near the surface.

Very wet soil
If rain has been falling for some time, the soil may become full of water. Any further rain is unable to soak into the ground and remains on the surface.

Very dry soil
Soil that is baked hard by the sun in dry weather builds up a crust. Rain is unable to soak through the crust and so remains on the surface.

Steep slopes
Rain falling on a steep slope runs quickly downhill towards a river. It has little time to soak into the ground, so most stays on the surface.

Floods are more common now than they used to be. There are more of them and they are increasing in size. Many people are blaming human activity for this.

Two ways in which humans may increase the risk of flooding are by cutting down trees and building more towns and cities. These are shown in drawings **C** and **D**.

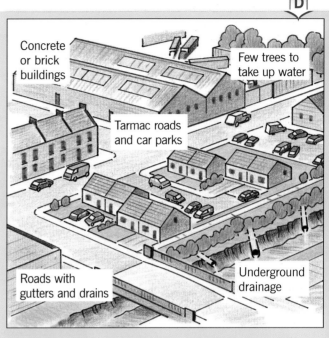

C

Leaves act as an umbrella and stop rain from hitting the ground

Water on leaves evaporates and does not reach the ground

Roots take up water and reduce the amount reaching the river

Roots slow down the movement of water in the soil

Cutting down trees (deforestation)

Many of the world's forests are being cleared to make way for other developments. In some countries the number of serious floods has more than doubled since large-scale tree clearing began.

D

Concrete or brick buildings

Few trees to take up water

Tarmac roads and car parks

Roads with gutters and drains

Underground drainage

Buildings and roads (urbanisation)

Rain falling on concrete and tarmac is unable to soak into the ground, so stays on the surface. Gutters and drains then carry the water quickly and directly to the river. Large towns are most at risk.

Activities

1 a Make a larger copy of drawing **E**.

 b Add the following labels to your drawing to show how a river floods:
 - River level rises
 - Water quickly reaches river
 - River floods
 - Water runs over surface
 - Heavy rain falls
 - Rain soaks into ground

2 Describe four factors that increase the risk of flooding.

3 With the help of diagram **F**, describe how:
 a cutting down trees, and
 b building towns can make floods worse.

4 Now complete the activities on page 109.

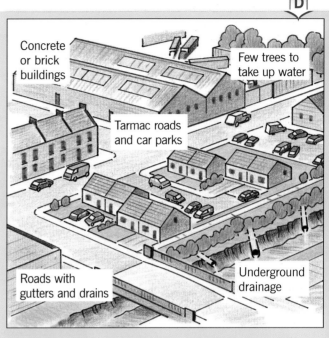

E

① ② ③ ④ ⑤ ⑥

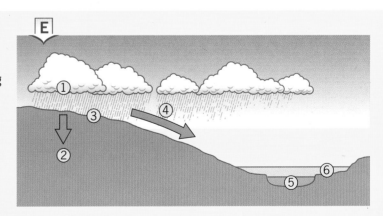

F

Summary

River flooding is most likely after heavy rain or rapid snow melt. The flood risk is greatest when water is unable to soak into the ground. Human activities can increase the chance of flooding.

York COURANT

Friday 28 September 2012

FLOODS HIT YORK AGAIN

Homeowners and businesses in the centre of York face another day of mopping up after one of the worst spells of flooding in its history.

More than a month's rain fell in 24 hours as torrential downpours and storms caused the River Ouse to burst its banks. River levels in York were the highest for 387 years as the Ouse reached a record high of 6.2 metres above normal.

Over a thousand properties were flooded and hundreds of people were forced to leave their homes. Villages to the south of York were also badly affected. Many people here have lost all their belongings, and councils have had to set up refugee centres at local schools and leisure centres.

Many roads were blocked and all rail services cancelled. Emergency services including lifeboats and coastguard helicopters helped rescue people trapped in buildings and on rooftops.

As people return to their homes, they are finding everything covered in a thick layer of foul-smelling mud. The clean-up operation will take weeks. Many people will not be back in their homes for months. Some businesses may never re-open.

Insurance experts are putting the cost at over £300 million but there are fears that many people will not be insured. The government has promised help to areas most in need.

For England and Wales, autumn 2012 was the wettest since records began in 1766. Major flooding affected large parts of the country. The area around York has always been liable to flood. In recent times, however, flooding has occurred more often and has been more serious than in the past. Many people blame human activity for this.

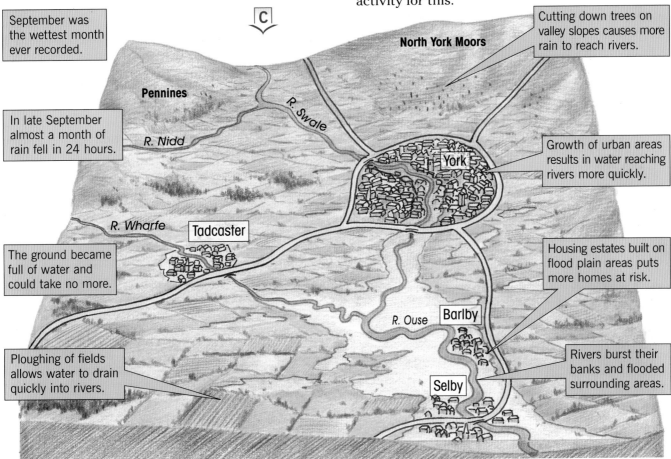

September was the wettest month ever recorded.

In late September almost a month of rain fell in 24 hours.

The ground became full of water and could take no more.

Ploughing of fields allows water to drain quickly into rivers.

Cutting down trees on valley slopes causes more rain to reach rivers.

Growth of urban areas results in water reaching rivers more quickly.

Housing estates built on flood plain areas puts more homes at risk.

Rivers burst their banks and flooded surrounding areas.

North York Moors

Pennines

R. Swale

R. Nidd

York

R. Wharfe

Tadcaster

R. Ouse

Barlby

Selby

Activities

1 a When did the flooding around York happen?
   b Which river flows through York?
   c Which three other rivers drain the area?
   d Which settlements were affected by flooding?
   e What are the names of the two upland areas?

2 Make a larger copy of drawing **D** and add six causes of the 2012 flood.

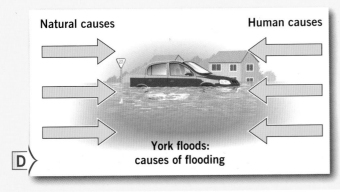

Natural causes   Human causes

York floods: causes of flooding

Record flood levels swamp York

Families hit as floods wreck homes

Problems continue as flood waters recede

Flood havoc hits commuters

3 Describe the effects of flooding in York using the newspaper headlines shown in **E**.

4 Write a description of the flood scenes shown in photos **A** and **B**.

Summary

Floods can cause much damage and seriously affect people's lives. There are usually several different causes of floods but some places are more at risk from floods than others.

How does the UK cope with floods?

Flooding is a serious problem in the UK and it is happening more often. There are currently over 5 million people in England and Wales who live and work in properties that are at risk of flooding from rivers or the sea. Autumn 2012 was the wettest since records began in 1766. Major flooding affected large parts of the country, and in some cases water levels were at their highest for 100 years. Whilst flooding cannot be prevented, in countries like Britain much can be done to reduce the risk of floods and limit their worst effects.

The **Environment Agency** is an organisation that looks after flooding from rivers and the sea in England. It monitors rainfall, river levels and sea conditions 24 hours a day. This information is used to predict the possibility of flooding. The Environment Agency's Floodline issues warnings. It also gives advice on what to do before, during and after a flood.

The Scottish Environment Protection Agency (SEPA), Natural Resources Wales and the Northern Ireland Rivers Authority are responsible for Scotland, Wales and Northern Ireland respectively.

A Planning for flooding in the UK

1. Study the UK's rivers and coasts and identify areas most at risk and where flooding would do most damage.

2. Recommend the building of flood defences such as embankments and overflow channels where they are needed.

3. Continually check rainfall and water levels to see if a river is going to flood.

4. When floods are expected, warn those in most danger by radio, TV, telephone, home visits, Twitter, Facebook and the internet.

5. Alert emergency services such as the police, fire brigade and army, to provide help for those in need.

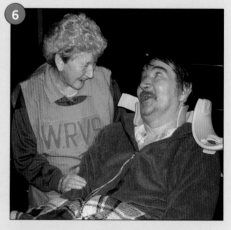

6. Ensure that food and shelter is available for those made homeless. Emergency medical care should also be available.

Flood warning codes

FLOOD ALERT

Flood Alert means flooding is possible. Be prepared.

FLOOD WARNING

Flood Warning means flooding of homes, businesses and main roads is expected. Immediate action required.

SEVERE FLOOD WARNING

Severe Flood Warning means severe flooding is expected with a danger to life and property. Immediate action required.

What to do in a flood

Before a flood

Be alert for flood warnings and take action.
Check on family and nearby neighbours.
Move people, pets and valuables to safety.
Collect warm clothes, food and a torch.
Block doorways with sandbags.
Switch-off electricity and gas.

During a flood

Listen to the local radio for flood news.
Never walk, drive or swim through flood water.
Avoid flood water as it may be contaminated.

After a flood

Check if it is safe to turn electricity and gas on.
Open windows and doors for ventilation.
Throw out contaminated food.
Wash taps and run them before use.
Clear up by disinfecting walls and floors.
Beware of rogue traders offering to help.
Call your insurance company for advice.

Activities

1. a Describe three ways that the Environment Agency can help reduce the risk of flooding.
   b Describe three ways that the Agency can help limit the worst effects of flooding.

2. Which flood warning would have been given for the York floods of September 2012 (pages 38 and 39)? Give reasons for your answer.

3. 'Floodline' encourages people to make a family flood plan like the one opposite. Write out the plan and add a reason for each point.

4. a Find out more about Floodline and the Environment Agency by visiting www.gov.uk/environment-agency.
   b Design a leaflet to give to people living in areas where there is a flood risk. Explain to them briefly what information is available and what they should do.

Family Flood Plan

- Know how to contact each other.
- Put together an emergency flood kit.
- Know how to turn off power supplies.
- Put emergency numbers in a safe place.
- Understand the flood warning system.
- Listen to the local radio programme.

Summary

There is no easy way to cope with floods. Rich countries like the UK can afford schemes that help reduce the damaging effects of flooding.

How can the risk of flooding be reduced?

There are many different ways of controlling rivers and reducing the risk of flooding. The methods shown below are called **flood prevention schemes** because they try to stop floods happening.

Many people now believe that complete river and flood control is impossible. They say that flooding should be allowed to happen as a natural event. Flood prevention schemes can, in the long term, save money. They also improve water quality and help support wildlife.

A

Dams
A dam built across a river traps water and stores it in a reservoir. The water may then be released in a controlled way.

Forests
Trees may be planted in the drainage basin. These will slow down water movement and reduce the amount reaching the river.

Embankments
The river's banks may be built up with earth or concrete. This will make the river deeper and keep the water in.

Concrete linings
Line river channels in urban areas with concrete. This will take excess water quickly away from danger areas.

Activities

1 Draw a star diagram to show eight ways of reducing the risk of flooding. Write a short sentence to describe each one.

2 Look at the different approaches to flood prevention. Which approach do you think:

  a costs most

  b costs least

  c may drown farmland and houses

  d uses up most land

  e protects the natural environment?

Give reasons for your answers.

3 One approach to flooding is simply to allow rivers to flood naturally. For each of the people below say if they would be **for** or **against** this method. Give reasons for your answer.

Local farmer   Flood protection manager   Bird watcher

Summary

A variety of methods can be used to reduce the risk of floods, but there is no way to stop flooding completely. A modern approach is to allow parts of a river to flood naturally.

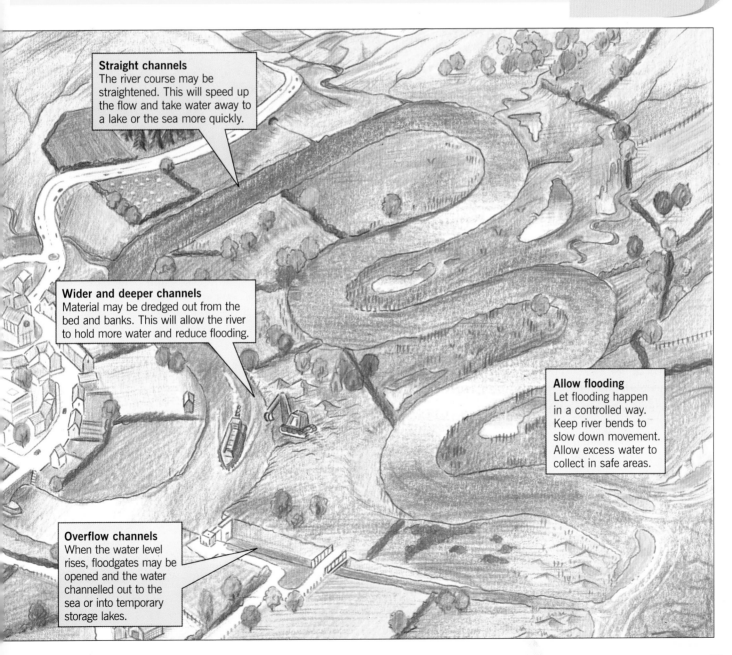

Straight channels
The river course may be straightened. This will speed up the flow and take water away to a lake or the sea more quickly.

Wider and deeper channels
Material may be dredged out from the bed and banks. This will allow the river to hold more water and reduce flooding.

Allow flooding
Let flooding happen in a controlled way. Keep river bends to slow down movement. Allow excess water to collect in safe areas.

Overflow channels
When the water level rises, floodgates may be opened and the water channelled out to the sea or into temporary storage lakes.

How should the Doveton valley be protected from flooding?

Look at map **D**, which shows part of the Doveton valley. In most years, the river overflows its banks and causes serious damage. The people in the area want their homes and land to be protected from the flooding.

The Environment Agency has agreed to look at the problem. It has made a study of the area and suggested four different schemes to help stop the flooding. It is your task to decide which is the best scheme.

Factors to consider	Scheme A	Scheme B	Scheme C	Scheme D
Prevents all flooding				
Stops flooding in Crofton				
No homes lost				
No roads submerged				
No grazing land lost				
No good farmland lost				
Helps with irrigation				
Helps protect wildlife				
Not too expensive				
Total				

A

B

1 a Copy table **A** which shows some factors that have to be considered when choosing a flood protection scheme.

  b Look carefully at the map and scheme descriptions. Show the advantages of each scheme by putting ticks in columns **A**, **B**, **C** or **D**. Complete one factor at a time. More than one column may be ticked for each factor.

  c Add up the ticks to find which scheme has the most advantages.

  d Which scheme would *you* choose?
  The one with the most advantages would be the best. If two schemes are equal, think about which parts of the valley you would want to protect most.

  e Briefly describe the scheme you have chosen. Explain how it will help protect the valley from flooding.

2 The flood protection scheme will affect different people in different ways. Work in pairs and discuss what the people in the drawing below will think of your chosen scheme. For each person say if they would be **for** or **against** the scheme. Give reasons for their views.

C

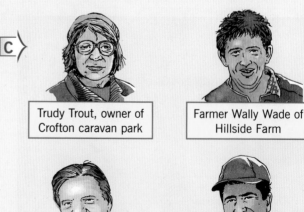

Trudy Trout, owner of Crofton caravan park

Farmer Wally Wade of Hillside Farm

Barry Beer, owner of the Crofton Inn

Larry Laugh, local lorry driver

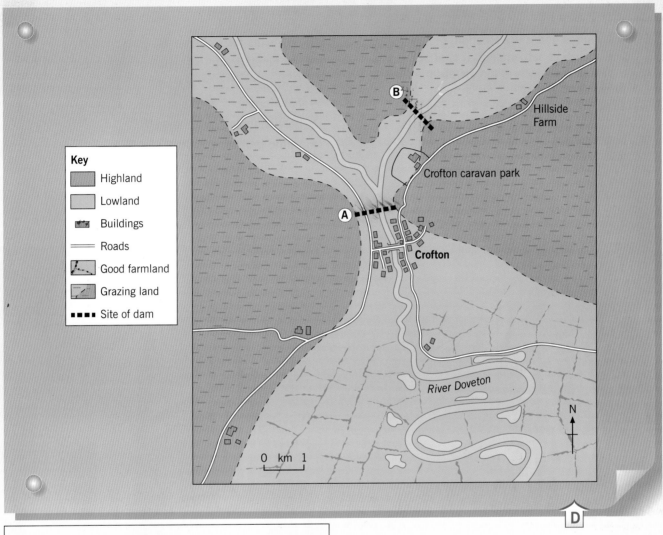

Key
- Highland
- Lowland
- Buildings
- Roads
- Good farmland
- Grazing land
- ▪▪▪▪ Site of dam

(Map labels: Hillside Farm, Crofton caravan park, Crofton, River Doveton, dam sites A and B, scale 0 km 1, N arrow, D)

Very expensive = **££££**   Quite cheap = **£**

Scheme A
Build a dam at A and create a large reservoir above the village. Much farmland and several farms would be flooded. The scheme would stop flooding in the village and protect most of the valley.   **Cost = ££££**

Scheme B
Build a dam at B and create a small reservoir higher up the valley. Nobody would lose their home but some grazing land used by sheep and cattle would be lost. There would still be some flooding in Crofton and further downstream.   **Cost = ££**

Scheme C
Build a dam at B and deepen the river channel through Crofton. This would allow the water flowing through the village to move away more quickly. The scheme would protect Crofton but there may still be some flooding downstream.   **Cost = £££**

Scheme D
Build embankments at Crofton. Deepen and straighten the river below the village to take water away quickly. Allow natural flooding to happen downstream at the river bends. There would still be some flooding, especially upstream of Crofton.   **Cost = £**

4

What are settlements like?

What is this unit about?

This unit is about the location, growth and nature of settlements. It is about **urbanisation** which is the increase in the proportion of people living in towns and cities. Urbanisation looks at the functions and land use of settlements and shows how changes may bring benefits but can also cause problems.

In this unit you will learn about:

- how sites for settlements were chosen
- the benefits and problems of settlement growth
- land use patterns in towns
- how functions and land use change
- how shopping has changed
- traffic problems and solutions
- how environments may be improved.

A   Central Birmingham, UK

Why is learning about settlements important?

Most of us live in a settlement of some kind. What settlements are like, therefore, affects us all in some way or another. This unit can help us understand what settlements are about and how they try to provide for the needs of people living in them. It can also help us see how changes to settlements can directly affect us, and have an impact on the way we live our lives.

This unit can also help you in other ways. It can:

- give you an interest in the place where you live
- help you choose where you might want to live in the future
- help you appreciate the problems facing town planners
- help you choose the best transport to use
- help you understand how the environment may be improved in the area where you live.

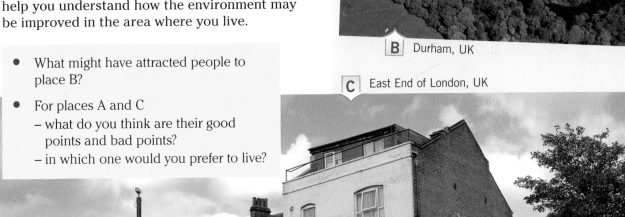

B Durham, UK

- What might have attracted people to place B?

- For places A and C
  - what do you think are their good points and bad points?
  - in which one would you prefer to live?

C East End of London, UK

47

How were the sites for early settlements chosen?

When we use the word **site** we mean the actual place where a village or town grew up. A site was chosen if it had one or more natural advantages.

Diagram **A** shows eight natural advantages. The more natural advantages a place had the more likely it was to grow in size.

Protection. Good views from a hilltop give you warning if you are about to be attacked.

Plenty of water. Needed for drinking, cooking and washing. Water might come from a river, spring or well.

Not too much water. Sites must not flood or be marshy.

Rivers. Easy to cross either on foot at a ford or by a bridge.

Building materials. Needed wood or stone. Useful to be near a wood or a rocky hillside.

Supply of wood. Needed for fires for warmth and to cook on.

Shelter. A south facing slope will have more sun and will be protected from the cold north wind.

Flat land. Easier to build on, for growing crops and travelling to other towns.

ROME RULES OK

UP THE BRITS

'Is this a good place to build a village?'

'Is this a good place to build a town?'

Activities

1 Write down the meaning of the word 'site'.

2 Landsketch **B** shows an area in Ancient Britain. On it, labelled A, B, C, D and E, are five possible sites for a village.

   a Suggest at least **one** natural advantage of each site.

   b Suggest at least **one** natural disadvantage of each site.

   c Which site would you choose? Give **three** reasons for your choice.

3 Try to find out what were the natural advantages of the site of your own town or village.

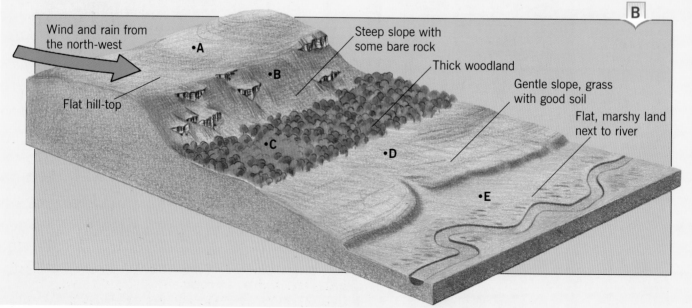

Wind and rain from the north-west

Flat hill-top

Steep slope with some bare rock

Thick woodland

Gentle slope, grass with good soil

Flat, marshy land next to river

•A   •B   •C   •D   •E

The photo below shows Warkworth, a village in Northumberland. It is located on a bend of the River Coquet. Early settlers were most concerned about their safety and getting a supply of food and water. As you can see, the site at Warkworth provided those needs.

Despite its good site, Warkworth has never grown into a large town. This is because the original advantages are no longer so important. Nowadays, people want to be near employment and services such as schools and hospitals. These are not so readily available at Warkworth.

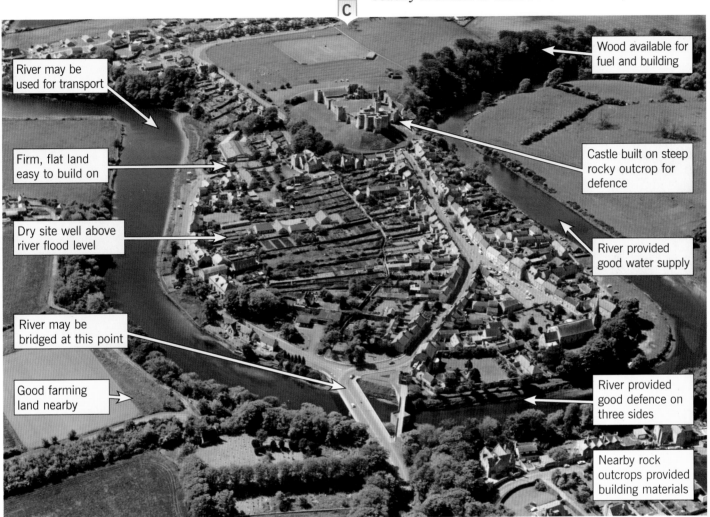

C

- River may be used for transport
- Firm, flat land easy to build on
- Dry site well above river flood level
- River may be bridged at this point
- Good farming land nearby
- Wood available for fuel and building
- Castle built on steep rocky outcrop for defence
- River provided good water supply
- River provided good defence on three sides
- Nearby rock outcrops provided building materials

Activities

1 Draw a star diagram like the one below to show the advantages of Warkworth as a site for a settlement. Give two advantages under each heading.

Defence — Site advantages of Warkworth — Food and water — Building materials — Building land

2 Complete table **D** to show how **some** of Warkworth's original site advantages are no longer so important.

D

Original advantage	Why no longer important
•	
•	

Summary

Early sites for settlements were chosen because of natural advantages such as a good water supply, dry land, defence, shelter, farmland and building materials.

What different settlement patterns are there?

If you look at map **D** on the next page you will see that the settlements have different shapes. Some are long and thin, some are compact and almost round, others are broken up and spread out. In geography we call these different shapes the **settlement pattern**.

Settlement patterns are usually influenced by the natural features of the area. These are often the same features that were considered important when choosing the original site for the settlement. The three main types of settlement pattern are shown below.

- A **dispersed settlement** has buildings that are well spread out.
- Settlements with this pattern are often found in highland areas where it is not easy to build houses close together. Here, people also needed more land to grow their crops or graze their animals.

Dispersed A

- A **nucleated settlement** has buildings closely grouped together.
- Settlements with this shape often grew around a road junction or river crossing. A long time ago people built their houses close together for safety. This pattern is common in lower, flatter parts of Britain.

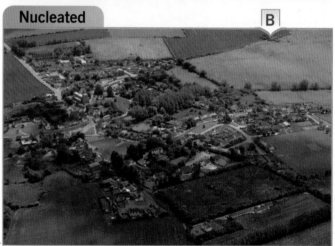

Nucleated B

- **Linear settlements** are often called **ribbon developments** because they have a long, narrow shape.
- Settlements with this shape usually grow along a narrow valley where there is little space. They may also be found strung along a road or on either side of a river.

Linear C

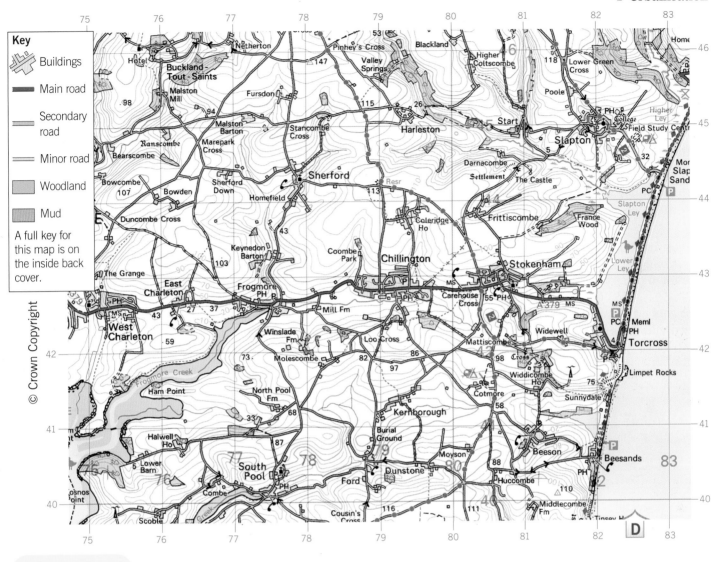

Key

Buildings

Main road

Secondary road

Minor road

Woodland

Mud

A full key for this map is on the inside back cover.

© Crown Copyright

Activities

1 Copy the settlement pattern drawings below.
Label each one **dispersed**, **nucleated** or **linear**.
Write a brief description of each one.
Suggest a reason for its shape. **E**

2 Map **D** is part of Devon in south-west England.
It shows many different settlement patterns.

a Make a larger copy of table **F**.

b Complete your table by filling in the empty boxes. The first one has been done to help you. (You may need to look at page 96 to remind you about grid references.)

c Find another example of a nucleated settlement and a linear settlement and add them to your table.

F

Village name	Map reference	Simple drawing	Settlement pattern
Bowden	7644	◆〜◆	Dispersed
Slapton	8144		
South Pool	7740		
Cotmore	8041		
Beeson			
Torcross			
Sherford			

Summary

The three main types of settlement pattern are dispersed, nucleated and linear. The shape of a settlement is usually determined by the physical features of the surrounding area.

How do settlements change with time?

No town or village remains the same for ever. Over a period of time the following may all change:

1 the **shape** of a settlement
2 the **function** of a settlement
3 the **land use** of a settlement
4 the **number** and **type** of people living in the settlement.

Villages are small in size so it is often easier to see these changes in them than it is to see changes in a large town or city.

What was a typical village like in the 1890s? Although no two villages are the same, most have several things in common. Diagram **A** shows a typical village about one hundred years ago. In the centre there was often a village green. Buildings were grouped closely together (nucleated) around this green forming a core. Roads were usually narrow lanes. Most houses were small terraced cottages. The people who lived in them would probably have been born in the village. Most would have worked on local farms. As houses and farms were built at different times they would have different styles and building materials.

How had the village changed by the 2010s? Diagram **B** shows the same village today. The village has grown larger and has many new buildings. It has become **suburbanised**. This means it has become similar to the outskirts of larger towns.

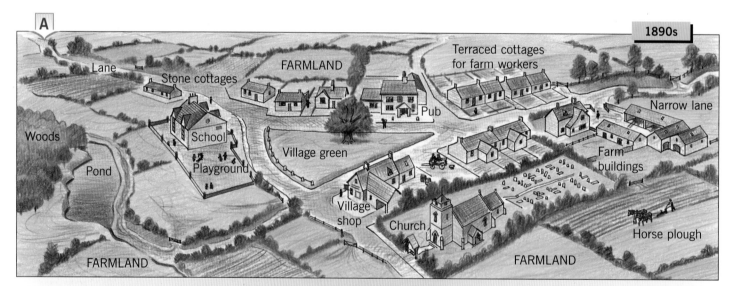

A   1890s

Lane · FARMLAND · Stone cottages · Terraced cottages for farm workers · Pub · Narrow lane · Woods · School · Village green · Farm buildings · Pond · Playground · Village shop · Church · Horse plough · FARMLAND · FARMLAND

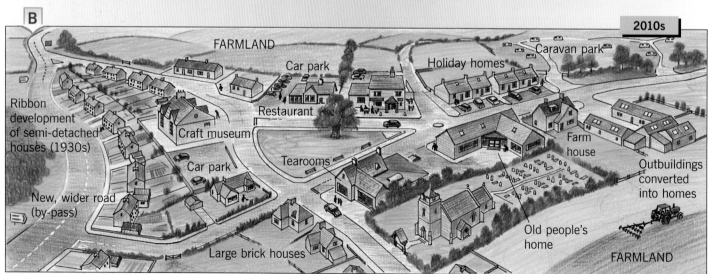

B   2010s

FARMLAND · Car park · Caravan park · Holiday homes · Ribbon development of semi-detached houses (1930s) · Restaurant · Craft museum · Farm house · Car park · Tearooms · Outbuildings converted into homes · New, wider road (by-pass) · Old people's home · Large brick houses · FARMLAND

Activities

1 Write down the meaning of:
   a   shape
   b   function
   c   land use
   when talking about a settlement.

2 **Spot the differences!** List at least **ten** differences between the village in the 1890s and the village in the 2010s.

3 The changes to the village will have affected different groups of people in different ways. Look at the pictures of some of these groups of people shown below. Match up the pictures with the statements below numbered **1** to **8**.

   For example:
   Young married couple = statement **2**

4 Activity **2** asked you to find the differences in the village between the 1890s and the 2010s. Why do you think changes have been made in:
   a   the number and type of houses
   b   the use of buildings around the green
   c   the use of the land around the village
   d   the roads?

5 It has been suggested that the woods should be cleared so that an estate of expensive houses can be built.
   a   Which groups of people will like this change?
   b   Which groups of people will be against this change?

   Suggest reasons for your answers.

How groups may be affected

1 I might have to close as most people have cars to shop in town.

2 We are just married and cannot afford an expensive house.

3 The extra noise frightens away the wildlife.

4 To get customers I have to provide food for townspeople. Villagers only want a drink.

5 I made money by selling my land so that houses could be built. Now people walk on the land I still own.

6 With all the new houses I have plenty of work to do.

7 I have to travel 10 km to school. At night there is nothing to do.

8 I came here for peace and quiet. Now I cannot drive into town and there are no buses.

Farmer

Shopkeeper

Bird watcher

Teenager

Young married couple

Restaurant owner

Elderly person

Builder

Summary

Settlements change over a period of time.
These changes can affect:
- the size and shape of the settlement
- the environment, e.g. new roads, larger villages
- the lives of people living in the settlement.

What are the benefits and problems of settlement growth?

In Britain most people are urban dwellers living in towns and cities. These settlements grew very quickly in the nineteenth century. This was when large numbers of people moved there to work. Today, Britain's cities are no longer growing in size. However, in many overseas countries people are still moving to cities in large numbers.

This is because they believe that many **benefits** come from living and working in cities. Moving there will improve their **quality of life**.

Drawing **A** shows some of the benefits which people hope to find in large towns and cities.

- There are more houses and flats to buy or to rent.
- There are more jobs which are better paid.
- Food supplies are more reliable, with many shops giving a greater choice.
- It takes less time and money to travel to work and to shops.
- There are more and better services, such as schools and hospitals.
- Urban areas have 'bright light' attractions, such as discos, concerts and sporting activities.

For people already living in cities, life is often less attractive. Living and working in cities creates many **problems**. Drawing **B** shows some of the problems found in cities.

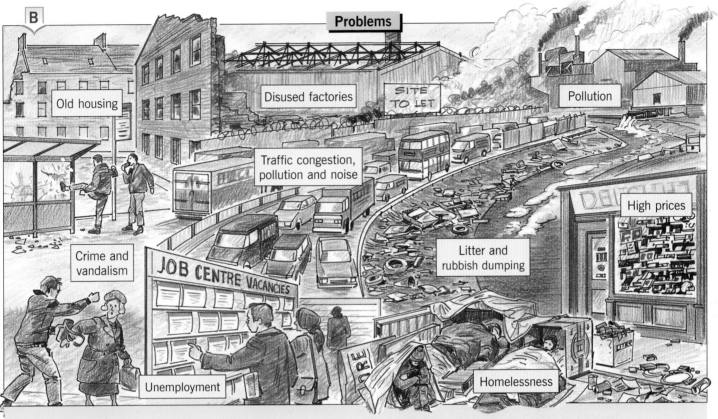

B

Problems

Old housing

Disused factories

SITE TO LET

Pollution

Traffic congestion, pollution and noise

High prices

Crime and vandalism

JOB CENTRE VACANCIES

Litter and rubbish dumping

Unemployment

Homelessness

- Traffic causes congestion, accidents, noise and air pollution.
- Old roads are too narrow for lorries and buses; new roads take up much land.
- Old houses and factories need urgent, expensive repairs or they are left empty.

- There is waste land where houses and factories have been pulled down.
- Crime, vandalism and litter make cities dangerous and unpleasant.
- Land is very expensive to buy, in and near the city centre.

Activities

1 a Make a copy of the table below. List the **three** things which you think are best about living in cities, and the **three** things you think are the worst.

Cities	
Good news	Bad news

  b Do you think there is more good news or bad news?

2 If you had to move from where you live, would it be to a bigger or a smaller settlement? Give reasons for your answer.

3 Try to find out what has been done in your local town or city to try to reduce:
  a traffic problems
  b pollution
  c crime, vandalism and litter.

4 Suggest other ways in which these three problems may be overcome.

Summary

Many people move to large cities because they see benefits in living and working there. However, as these settlements become older and bigger, many problems are created.

Why are there different land use patterns in towns?

When each town first began to grow it usually had one particular use or **function**. Towns and cities of today often have several different functions. The main functions are **commerce** (shops and offices), **industry** (factories), **residential** (flats and houses) and **open space** (parks and sports facilities). As each function tends to be found in a particular part of a town, then a pattern of land use develops. Although no two towns will have exactly the same pattern of land use, most have similar patterns. When a simple map is drawn to show these similarities it is called an **urban model**.

The model in diagram **A** shows four differences in land use drawn as a series of circles around the city centre. It is suggested that this pattern developed for two reasons:

1 The oldest part of a town is in the middle. As the town grew, larger new buildings were built on the edges.
2 Land in the city centre is expensive to buy. This is because many different types of land users would like this site and so they compete for it. Usually the price of land falls towards the edges of towns.

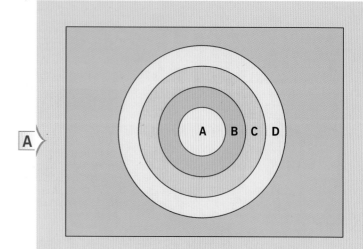

A

Zone A The central business district (CBD)
The centre of the town was the first place to be built. It is still full of shops, offices, banks and restaurants. There are very few houses and little open space here.

Zone B The inner city
This used to be full of large factories and rows of terraced housing built in the nineteenth century. Houses were small and there was no open space as land was expensive. Today most of the big factories have closed and the oldest houses have been replaced or modernised.

Zone C The inner suburbs
This is mainly semi-detached housing built in the 1920s and 1930s. There is some open space.

Zone D The outer suburbs
This includes large, modern houses and some council estates built since the 1970s. Recently small industrial estates, business parks and large supermarkets have been built here. There are large areas of open space.

Activities

1 a Copy diagram **A** and colour in the four zones.
  b Name the four zones.
  c Name an area in your local town for each of the zones.

2 Diagram **B** shows four houses that are for sale. In terms of location, cost and amenities, which house might you choose:
  a if you were a first-time home buyer
  b if you had two children aged under 6 years
  c when your children leave home and you have a good job
  d when it is time for you to retire from work
  e if you could choose for yourself as a teenager?

  In each case give reasons for your answer.

Summary
The main function, or land use, of an area may result from its age and the cost of land.

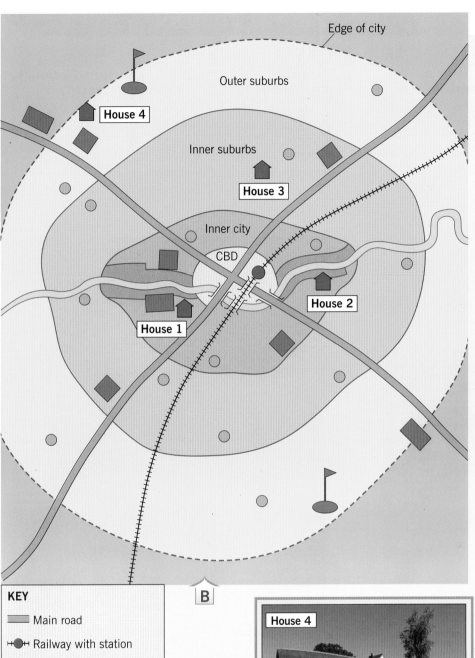

Edge of city

Outer suburbs

House 4

Inner suburbs

House 3

Inner city

CBD

House 2

House 1

B

KEY

- Main road
- Railway with station
- River
- House for sale
- Regional shopping centre
- Secondary school
- Primary school
- Golf course
- Old industrial area
- Countryside

House 1

Mid-terraced
- Kitchen
- Living room
- Two bedrooms
- Bathroom extension in backyard
- In need of some modernisation

£100,000

House 2

Modern fifth-floor apartment/flat in a recently converted factory
- Modern kitchen
- Living room with dining area
- Two bedrooms
- Bathroom
- Parking space

£180,000

House 4

Large detached house
- Kitchen and dining area
- Living-room with large conservatory
- Four bedrooms, one en-suite
- Family bathroom and downstairs WC
- Utility room/study
- Garage
- Spacious gardens at front and rear

£600,000

House 3

Semi-detached in a cul-de-sac
- Kitchen
- Living room and dining room
- Three bedrooms
- Bathroom and separate WC
- Garage and small conservatory
- Garden front and rear

£300,000

Why does land use in towns change?

Land use in towns changes over time. City centres are modernised to attract more people, while open space on the outskirts is turned into housing estates, large shopping centres and industrial parks.

Most **inner city** areas were built over a hundred years ago. Naturally they have aged in that time. Houses became too old and cramped to live in and factories closed down. The inner cities had to change.

Over the years, most towns have tried to improve conditions in the old inner city areas. London Docklands is an example of a scheme which has brought change and improvement to an area.

A

B

London Docklands – the problem

- Up to the early 1950s, London was the busiest port in the world and the Docklands a thriving industrial zone.
- For a number of reasons, shipping on the Thames went into decline, and by 1981 the docks were virtually abandoned and derelict.
- By that time there were very few jobs, transport was poor and there was a lack of basic services.
- Housing was a particular problem, with many old terraced houses lacking a bathroom or indoor toilet and in need of urgent repair.
- Although conditions were difficult, a strong 'EastEnders' community spirit built up.

London Docklands – the solution

- In 1981 the London Dockland Development Corporation was set up with the aim of improving living and working conditions.
- It began by clearing the old docks and houses and turning warehouses into expensive flats.
- Old industries were replaced with those using high-technology, such as newspapers, and by office blocks of financial firms.
- Underground stations were improved and a new City Airport and Docklands Light Railway built.
- The environment has been improved, trees planted and new parklands created.

How has the development affected people?

Many people have benefited from the Docklands redevelopment and are in favour of the scheme. Others, however, are less happy and are against it. Local people particularly feel disadvantaged. They say that housing is too expensive for them, that money has been spent on facilities for the rich rather than the poor, and that most jobs are inappropriate to their needs. They also think that the 'yuppie' newcomers rarely mix with local people and that the 'EastEnders' community spirit has been broken up.

C London Docklands – some winners and losers

Young married couple
We will have to move as we cannot afford to buy a place to live. A cheap flat is over £180,000.

Local shopkeepers
All these newcomers mean more trade – especially as they have plenty of money to spend.

Financial manager
We have modern offices and there is good-quality housing here. It only takes 10 minutes to travel into central London.

Local people
Most new jobs go to highly skilled people from outside the area. Our close-knit community has been broken up.

Elderly people
Shopping is expensive. Money is spent on houses and offices, not on hospitals and old people.

School leavers
We are familiar with the latest IT systems. We will be able to get jobs and stay in the area.

Activities

1 a What was the early success of Docklands based on?
   b What was the reason for its decline?
   c What problems were caused by this decline?
   d Describe a typical house shown in photo **A**.
   e What is meant by 'a good community spirit'?

2 Make a copy of diagram **D** below and use the headings to describe the changes to London Docklands.

Buildings

Industry/Jobs

Transport

Environment

D

3 Look carefully at drawing **C** above.
   a Which people do you think are winners?
   b Which people do you think are losers?
   Give reasons for your answers.

4 Overall, do you think the changes to London Docklands have been good or bad? Give reasons for your answer – but remember to think about different points of view.

Summary

As time passes, the functions and land uses of different parts of a town will change. These changes affect different groups of people in different ways.

Where do we shop?

We all go shopping. We need to buy things so that we can feed ourselves and live our lives. Some of us shop simply for enjoyment. Recent surveys have found that shopping is one of Britain's most popular leisure activities.

Shopping is also big business. Each year, people in the UK spend over £63 billion on food shopping alone. A further £84 billion is spent on clothes and shoes. More than 3.2 million people in the UK work in shops, many of them part-time.

As we have seen, smaller settlements usually have very few shops. Larger settlements, however, are likely to have several shopping centres. In recent years, many of these have been built out of town and well away from the city centre.

Where people choose to shop depends upon what they want to buy and how often they need that product. The larger the shopping centre the more choice of goods and services there will be. People will travel long distances to these centres.

- Some **goods** such as food and newspapers don't cost very much and may be needed every day. We are happy to buy them in the nearest convenient place.
- These are called **convenience** or **low order goods**. They may be bought at the local **corner shop**, nearby shopping centre or supermarket.

A

- Some goods like clothes and furniture are much more expensive. We buy them less often and like to compare styles and prices before we buy.
- These are called **comparison** or **high order goods**. They are bought at large shopping centres, either in the city centre or out of town. Here, there is usually a good choice and lower prices because of competition.

B

Activities

1 Complete these sentences.
   a Convenience goods are ...
   b Comparison goods are ...
   c A corner shop is ...

2 Look at the goods shown in drawing **C**. Sort them into two groups: convenience goods and comparison goods.

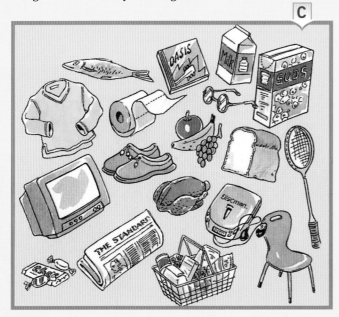

3 Now decide where you would buy the goods in drawing **C**. Look at drawing **D**, which gives you three choices. You must make a separate trip to each centre. Write out a shopping list for each visit.

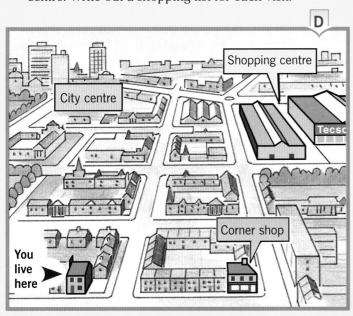

4 You may be close enough to a shopping centre to visit it during a geography lesson. If you do you **must** go in groups and take great care when crossing roads.

- **Aim** – to compare the shopping habits of people at your local shopping centre with those of people in the city centre.
- **Equipment** – questionnaire, clipboard and pencil.
- **Method**
   a Make up a questionnaire similar to the completed one below.
   b Politely ask at least 20 male and female shoppers of different ages the questions.
   c Share your answers with other groups.
   d Think of ways to illustrate your results.
   e Describe your findings.
   f Suggest reasons for differences between the results of your questionnaire and those of the one taken in the city centre.

Shopping survey

The numbers on the right show the result of asking 100 people in a city centre shopping area (mall) the following questions:

1 Do you shop here
   • every day? 15
   • two or three times a week? 15
   • once a week? 50
   • once a month? 20

2 Have you travelled
   • less than 1 mile? 15
   • between 1 and 2 miles? 20
   • between 2 and 5 miles? 30
   • over 5 miles? 35

3 Do you travel here
   • on foot? 5
   • by car? 75
   • by bus? 15
   • any other way? 5

4 Do you do most of your weekly shopping here?
   • Yes 75
   • No 25

5 What is the main thing you buy here?
   • Food 30
   • Clothes 40
   • Furniture 10
   • Domestic equipment 10
   • Others 10

Summary

There are many different types of shopping centre. The larger the centre, the greater the choice of shops and goods there are to buy.

How has shopping changed?

The city centre is the main shopping area in a town. It has the largest number of shops, the biggest shops and the most shoppers. People are willing to travel long distances to the city centre because of the great choice of goods that they can buy there.

The main advantage of the city centre is its **accessibility**. Most of the main roads, bus routes and rail systems from the suburbs and surrounding areas meet at the city centre. It is therefore the easiest place for most people living in the town to reach.

City centre shopping has changed a lot recently. Attempts have been made to reduce traffic congestion, and to provide for the safety and comfort of shoppers. Many towns now have covered **shopping malls** which give protection from the weather and are traffic free.

City centre

A

- Almost anything can be bought here. There are large department stores, nationwide supermarkets, chain stores and many specialist shops. Competition between shops keeps prices low.

- Good accessibility by car and public transport make it the most visited and busiest type of shopping centre.

- Overcrowding and traffic congestion cause problems. Pedestrianised areas and covered walkways have improved shopper comfort.

Activities

1 a Name the main street, shopping area or mall in your local town or city.
   b Name one department store, one nationwide supermarket and three specialist shops found in it.

2 Make a larger copy of sketch **B** showing a city centre shopping centre. (Your sketch does not have to be as detailed as sketch **B**).
   Complete it by answering the questions.

 B

f What is being done to improve city centre shopping?

e What are the main problems of these centres?

a What makes the city centre the main shopping centre?

b Why are people willing to travel long distances to this centre?

c Name four different types of shops in the city centre.

d What three things make the city centre accessible?

The biggest change in shopping has been the development of huge out-of-town shopping centres. These are located on the edge of cities, usually next to a main road or motorway. They are designed to attract motorists from a wide area and offer good accessibility and free parking.

Many people are worried about the development of these centres. They are concerned that they take trade away from the traditional shopping outlets. Some city centres have lost up to half of their business in the last ten years and are in serious decline. Many smaller shopping parades and corner shops have had to close altogether.

Another drawback of out-of-town shopping is the increased use of cars. This has caused more air pollution, noise and traffic congestion in suburban areas.

The government is very concerned about these effects. In future, it might not give permission to build any more out-of-town centres.

Out of town

- Ideal for shopping by car, with good road access and free car parking.

- Contain a large number of shops with a wide choice of goods. Prices are kept low by bulk buying and low running costs.

- Popular with shoppers who enjoy the bright and attractive air-conditioned shopping malls. Security staff ensure safety for families.

- Most centres have cafés, restaurants, cinemas and a wide range of other leisure amenities.

C

3 Design a poster to show the advantages of using a shopping centre like the one shown in photo C. Your poster should be attractive and interesting, and show the facts.

D

4 Imagine that you own a small shop close to a new out-of-town shopping centre. Your profits are down and you think you may soon have to close. Write a letter to the local council to say that the centre was a mistake. Mention all the ways that you think it is harming the area.

Summary

Shopping habits are changing. The city centre has always been the main shopping area in a town but it is now often congested and expensive. As more people shop by car, modern out-of-town centres are becoming increasingly popular.

Traffic in urban areas – why is it a problem?

Traffic is a serious problem in most urban areas. Large cities such as London, Paris and New York have over a million cars trying to move around in their central areas. Movement is often impossible. Perhaps worse than **congestion** is the problem of pollution.

Exhaust fumes are poisonous and can seriously damage health. Some city workers are so concerned that they wear masks to protect themselves. So what can be done? What are the causes of the problem, and why have they not been solved?

What is the problem?

Look at any urban area and you will soon be able to answer this question. Cars, buses and lorries all over the place cause congestion and chaos. They produce a lot of fumes, noise and danger. Other effects are:

- traffic jams blocking roads and stopping all movement
- delays for police, fire service and ambulances
- slow movement of people and goods
- loss of business and money
- people and buildings affected by noise and vibrations
- danger from accidents
- harmful exhaust fumes
- lack of parking places.

We live in an age of rapid transport yet vehicle movement is now actually slower than it was 80 years ago.

Activities

1 Look at the information with photo **A**. List what you think are the five worst problems caused by increased traffic in towns.

2 The two people in **B** are badly affected by traffic congestion and pollution. For each person write a letter to the local MP explaining how the problem affects them personally.

A businessman who lives out of town and drives to the city centre each day to work

A local resident with two young children who lives close to the main road

What is the cause?

There are many reasons for the traffic problems in our cities. The main one is simply that the number of cars has increased at a tremendous rate and there are now too many cars for cities to handle. It is predicted that this increase in cars will continue. By 2025 the number of cars might double and the number of lorries be three times greater than in 2010.

Another reason for traffic problems in cities is that most city centres were designed and built before cars were invented. They are therefore just not suited to today's transport. The problem is worst in the morning and in the late afternoon when people are travelling to and from work. This is called **the rush hour**.

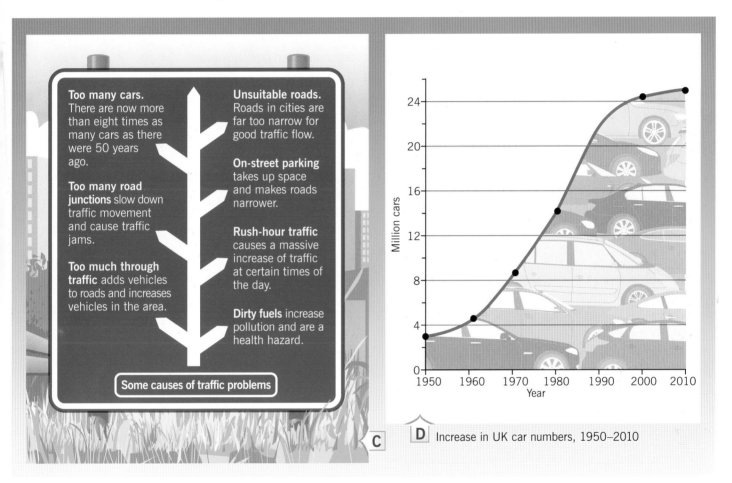

Too many cars. There are now more than eight times as many cars as there were 50 years ago.

Too many road junctions slow down traffic movement and cause traffic jams.

Too much through traffic adds vehicles to roads and increases vehicles in the area.

Unsuitable roads. Roads in cities are far too narrow for good traffic flow.

On-street parking takes up space and makes roads narrower.

Rush-hour traffic causes a massive increase of traffic at certain times of the day.

Dirty fuels increase pollution and are a health hazard.

Some causes of traffic problems

C

D Increase in UK car numbers, 1950–2010

3 Look at drawing **C**. List what you think are the five worst problems caused by increased traffic in cities. You need only write out the words in **bold**.

4 Look at graph **D**.
   a How many cars were on Britain's roads in 1950? How many were there in 2010?
   b Which of the graphs in **E** looks most like graph **D**? Use that description to describe the change in car numbers between 1950 and 2010.

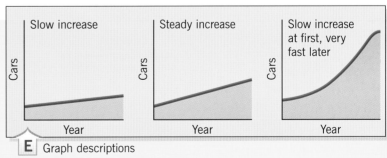

E Graph descriptions

Summary

Congestion and pollution are major problems in urban areas. The main causes of these problems are too many cars, rush-hour traffic and unsuitable roads.

Traffic in urban areas – is there a solution?

There are two main ways of approaching the problem. The first is to allow **private transport** to increase and make improvements to cope with larger amounts of traffic. The second is to restrict private transport and discourage motorists from bringing cars into town centres. This would mean improving **public transport** such as bus and train services.

In fact, the traffic problem is so big and complicated that no single solution will ever completely solve it. The best way is to try to reduce the worst parts of the problem by using several solutions together. Some ideas that have been tried are shown in diagram **A**. Can you think of any others?

A

Encourage private transport

Better traffic management:
• roundabouts
• one-way systems
• traffic lights

More off-street parking

Urban motorways to improve traffic flow

By-passes to keep through traffic out of towns

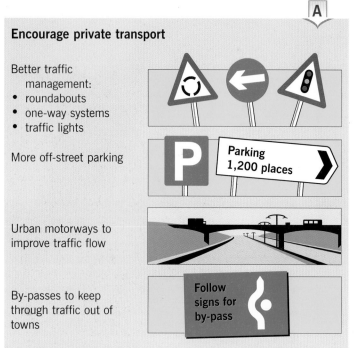

Discourage private transport

Improve public transport:
• reduce fares
• increase speed and comfort
• develop rail routes like the London Underground

Park-and-ride schemes:
• leave cars on town outskirts
• travel by free bus to centre

Make car parking difficult:
• increase charges
• reduce spaces

Congestion charging:
• motorists have to pay to use busy areas

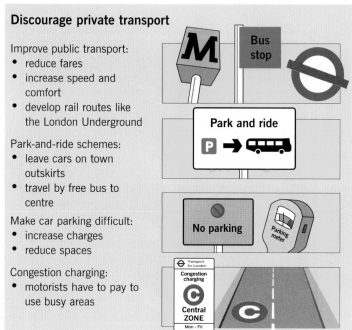

Activities

1 a What is meant by public transport?
   b What is meant by private transport?
   c Name each of the following types of transport and sort them into **Public** and **Private**.

2 Draw a poster to discourage motorists from taking their cars into town centres.

   • Show the bad things about town centres.

   • Show the other types of transport that can be used.

   • Colour your poster and make it interesting and attractive.

3 Diagram **B** shows how private and public transport in towns affect each other.

   Copy and complete the diagram using the following phrases.

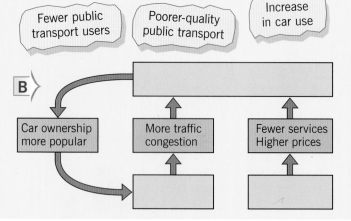

Public transport systems

In an attempt to reduce congestion and pollution, some cities have built new public transport systems. The aim of these systems is to move people around as safely, quickly and cheaply as possible. They are also designed to reduce pollution and so protect the environment.

Manchester's Metrolink tram system is an example of a surface light railway. This means that it runs on rails along existing roads and sections of former railways. Priority is given to the tram at traffic lights and road crossings. Travel by Metrolink has proven faster and cheaper to use than car transport.

The Manchester Metrolink C

- Connects suburbs to city centre
- Provides convenient way to travel in city centre
- Links with existing bus and rail routes
- Trams every 6 minutes during daytime
- Each tram can carry 206 passengers
- Fares subsidised to provide cheap travel
- Automatic ticket machines reduce costs

- Speeds up to 80 km/hr on former railways
- 26 stations and over 29 km of track
- Excellent wheelchair and pram access
- Powered by electricity so reduces noise and air pollution
- Carries over 16 million passengers a year
- Takes up to 2.5 million car journeys a year off the road

4 How does the Metrolink:
  a provide a fast and cheap service
  b serve people living away from the system
  c help protect the environment?

5 Public transport is not popular with everyone. Make a list of the disadvantages of a system such as Metrolink.

6 Describe a traffic problem near your school or where you live. Suggest how the problem could be reduced.

Summary

Solving the problem of urban traffic is difficult. Better public transport may be the best way to improve people's movements without further damaging the environment.

Where should the by-pass go?

When a place becomes too crowded with vehicles (congested), a road can be built around it to take away some of the traffic. A road that is built to avoid a congested area is called a **by-pass**. Some by-passes are very long. The M25 which goes all the way round London is over 160 km (100 miles) long. Most by-passes are much shorter than this.

Activities

Look carefully at drawing **C** opposite. It shows the area around Haydon Bridge, a small town on the banks of the River Tyne between Newcastle and Carlisle.

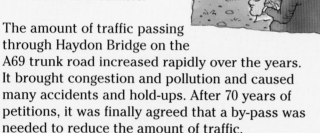

The amount of traffic passing through Haydon Bridge on the A69 trunk road increased rapidly over the years. It brought congestion and pollution and caused many accidents and hold-ups. After 70 years of petitions, it was finally agreed that a by-pass was needed to reduce the amount of traffic.

Of the three possible routes suggested, the red route was eventually chosen. The new by-pass was opened in 2009.

1 a Copy table **A**, which shows some things that should be considered when choosing a by-pass route.

   b Show the advantages of each route by putting ticks in the **Red, Blue** or **Yellow** columns. More than one column may be ticked for each point.

   c Add up the ticks to find out which route has the most advantages.

2 a Describe the red route. Start with: 'The by-pass begins at …'

   b Give three reasons why this route was chosen.

   c Give two advantages of this route.

Building a by-pass is not easy. Money has to be found, suitable routes planned out, and discussions held between people whom the route may affect. This is all very difficult and takes a long time.

Considerations	Red route	Blue route	Yellow route
Is the shortest route			
Avoids all the built-up area			
Avoids best farmland			
Avoids steep slopes			
Avoids floodplain			
Avoids caravan park			
Needs fewest bridges			
Requires fewest trees to be cut down			
Avoids new housing estate			
Avoids sports park			
Total			

A

3 Work in pairs and suggest which of the people in **B** would have been against:

   a the blue route

   b the yellow route.

Give reasons for your answer.

Walkers

Factory managers on industrial estate

Residents of Peelwell

Hotel owner in Haydon Bridge

B

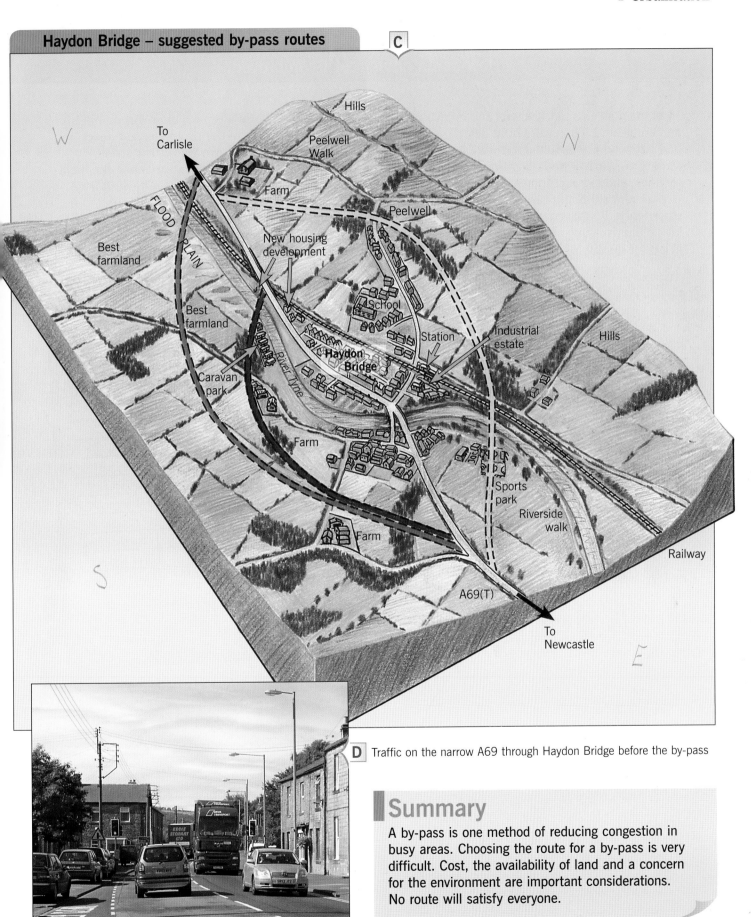

Haydon Bridge – suggested by-pass routes

C

Hills

Peelwell Walk

To Carlisle

W

N

Farm

FLOOD PLAIN

Peelwell

Best farmland

New housing development

Best farmland

School

Station

Industrial estate

Hills

Caravan park

River Tyne

Haydon Bridge

Farm

Sports park

Riverside walk

Farm

Railway

S

A69(T)

To Newcastle

E

D Traffic on the narrow A69 through Haydon Bridge before the by-pass

Summary

A by-pass is one method of reducing congestion in busy areas. Choosing the route for a by-pass is very difficult. Cost, the availability of land and a concern for the environment are important considerations. No route will satisfy everyone.

The urbanisation enquiry

How can a city street be improved?

Most main streets in UK towns and cities are busy and congested. They were built before people had cars and are unsuited to today's needs. Many streets are narrow and dangerous, and polluted by noise and exhaust fumes. There is no room to park and shopping can be an unpleasant experience. Buildings are often ugly and there is a lack of landscaping and open space. Overall, there is a poor **quality of environment**.

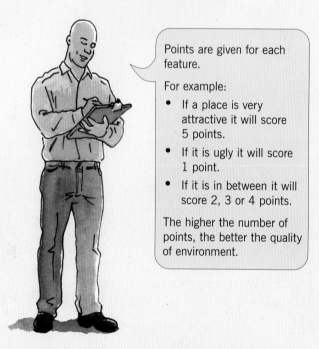

Points are given for each feature.

For example:

- If a place is very attractive it will score 5 points.
- If it is ugly it will score 1 point.
- If it is in between it will score 2, 3 or 4 points.

The higher the number of points, the better the quality of environment.

Drawing **A** shows a typical main street in a UK town with many of the problems just described. The people in the town want their street improved, and they have approached the local authority with their concerns. The authority agreed to produce a scheme that would:

1 reduce traffic congestion and pollution
2 make the area safer and more attractive
3 improve local shopping facilities.

QUALITY OF ENVIRONMENT SURVEY SHEET

	High quality 5 4 3 2 1					Low quality
Attractive						Ugly
Quiet						Noisy
Tidy						Untidy
Safe						Dangerous
Few cars						Many cars
Easy movement						Congested
Good shopping						Poor shopping
Good parking						Poor parking
Open space						No open space
Like						Dislike

Place .... Total .... out of 50

Drawing **B** shows the proposed improvement scheme. This must be discussed, and alterations suggested, before building can begin.

1 Make a list of the problems shown in drawing **A**.

2 Now look carefully at the improvement scheme in drawing **B**.
  a How has traffic congestion been reduced?
  b What has been done to improve safety?
  c What has been done to make the area more attractive?
  d How has shopping been improved?

3 a Make two copies of the survey sheet on the right.
  b Complete a survey for drawing **A**.
  c Tick the points you would give for each feature and add up the total number of points.
  d Complete a similar survey for drawing **B**.

4 Use the two surveys to measure the success of the improvement scheme. What features still need to be improved? Suggest what could be done to make these better.

5 The views of local people must be considered. For each person in **C**, say if they would be **for** or **against** the scheme. Give reasons for their views.

6 Should the scheme go ahead? Write a letter to the local authority giving your views and suggestions.

A — The area before improvements

B — Proposed improvement scheme

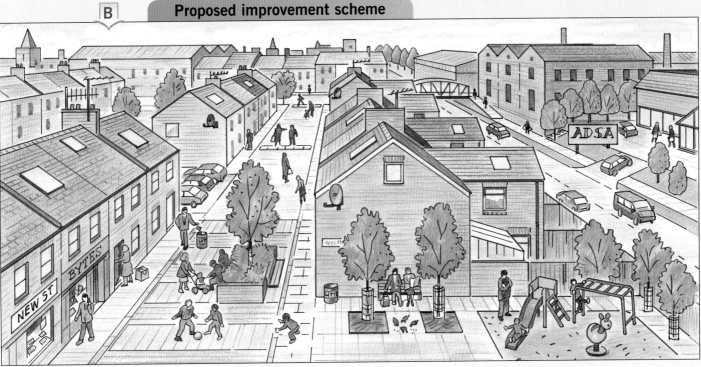

C

Mrs Briggs
A mother with two young children living on Main Street.

Mr Banks
Owner of a shop on Main Street. Often has to drive to his other shops in nearby towns.

Mr and Mrs Bell
Owners of a shop on Main Street which will be knocked down for parking.

5 What is Kenya like?

What is this unit about?

After a general introduction on Africa, this unit is about Kenya, a developing country. Although it is very poor economically, it is rich in scenery and wildlife and has cheerful people and some great athletes.

In this unit you will learn about:

- Africa's main features
- Kenya's main physical features
- population distribution and movement
- differences in urban and rural life
- the features of a developing country.

Why is learning about Kenya important?

Learning about Kenya gives you an interest and knowledge of people and places that are very different from those found in the UK.

This unit will help you to:

- broaden your knowledge of our world
- learn about different landscapes and climate
- understand ways of life that are different from your own
- recognise differences in a country
- develop an interest in other countries
- understand differences in development.

A   Mt Kilimanjaro from Amboseli National Park, Kenya

B Nairobi, Kenya

- Compared with the area where you live, how different is:
  – the countryside in photo **A**
  – the city in photo **B**
  – the village in photo **C**?

- Make a list of at least six words to describe the Kenyans shown in photo **D**.

C Samburu village, Kenya

D Kenyan schoolchildren

What are Africa's main physical features?

A   Sahara Desert, Algeria

The map of Africa **D** opposite is an enhanced satellite image. The colours have been changed slightly to make the features clearer. If you look carefully you should see the main mountain ranges, the rainforests that show as green, and drier areas that are yellow or brown. Pages 116 and 117 show how satellite images may be used in geography.

Africa is the world's second largest continent. It has a great variety of physical features including huge deserts, vast areas of tropical rainforest and extensive mountain ranges. Temperatures vary from over 40 °C in the deserts to –20 °C in the mountain areas. Three of the world's greatest rivers, the Nile, Congo and Zambezi, are also located in Africa.

Activities

1   Use the scale-line to measure the following. Page 94 will help you.
   a   The length of Africa from (A) to (B)
   b   The width of Africa from (C) to (D)
   c   The length of the rivers Congo and Zambezi
   d   The total length of the Great Rift Valley

2   Give the latitude and longitude of each of the following. Pages 10 and 11 will help you.
   a   Lake Victoria
   b   The mouth of the Nile
   c   The Cape of Good Hope
   d   The Atlas mountains
   e   Victoria Falls

3   Describe Africa's physical features using the headings below. Write a few short statements for each one.

   Africa: physical features
   • Earthquakes and volcanic activity
   • Vegetation features
   • River features
   • Wildlife

4   Write short descriptions of photographs **A**, **B** and **C**. Pages 110 and 111 will help you.

B   Victoria Falls, Zambia/Zimbabwe

C   Okavango Delta, Botswana

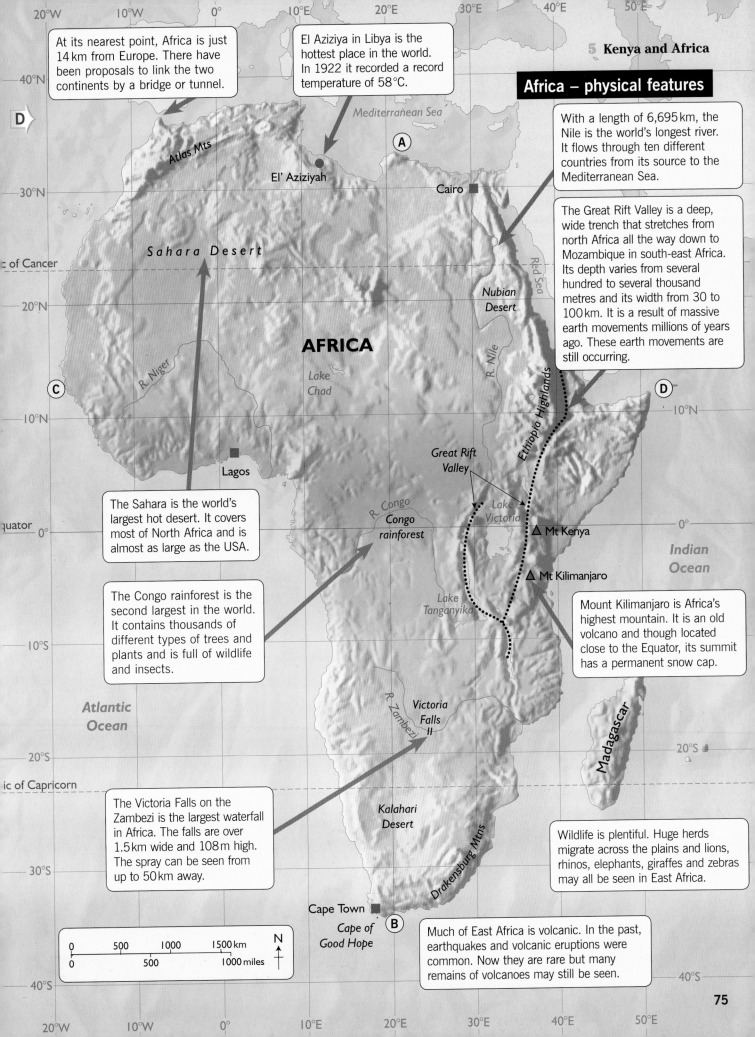

Africa – physical features

At its nearest point, Africa is just 14 km from Europe. There have been proposals to link the two continents by a bridge or tunnel.

El Aziziya in Libya is the hottest place in the world. In 1922 it recorded a record temperature of 58°C.

With a length of 6,695 km, the Nile is the world's longest river. It flows through ten different countries from its source to the Mediterranean Sea.

The Great Rift Valley is a deep, wide trench that stretches from north Africa all the way down to Mozambique in south-east Africa. Its depth varies from several hundred to several thousand metres and its width from 30 to 100 km. It is a result of massive earth movements millions of years ago. These earth movements are still occurring.

The Sahara is the world's largest hot desert. It covers most of North Africa and is almost as large as the USA.

The Congo rainforest is the second largest in the world. It contains thousands of different types of trees and plants and is full of wildlife and insects.

Mount Kilimanjaro is Africa's highest mountain. It is an old volcano and though located close to the Equator, its summit has a permanent snow cap.

The Victoria Falls on the Zambezi is the largest waterfall in Africa. The falls are over 1.5 km wide and 108 m high. The spray can be seen from up to 50 km away.

Wildlife is plentiful. Huge herds migrate across the plains and lions, rhinos, elephants, giraffes and zebras may all be seen in East Africa.

Much of East Africa is volcanic. In the past, earthquakes and volcanic eruptions were common. Now they are rare but many remains of volcanoes may still be seen.

Mediterranean Sea

Atlas Mts

El' Aziziyah

Cairo

Sahara Desert

c of Cancer

Nubian Desert

Red Sea

R. Niger

AFRICA

Lake Chad

R. Nile

Ethiopia Highlands

Lagos

Great Rift Valley

Lake Victoria

△ Mt Kenya

quator

R. Congo

Congo rainforest

△ Mt Kilimanjaro

Indian Ocean

Lake Tanganyika

Atlantic Ocean

R. Zambezi

Victoria Falls

Kalahari Desert

Madagascar

ic of Capricorn

Drakensburg Mtns

Cape Town

Cape of Good Hope

| 0 | 500 | 1000 | 1500 km |
| 0 | | 500 | 1000 miles |

N

What are Africa's main human features?

A Tahrir Square, Cairo, Egypt

With a total population of over a billion, Africa is the world's second most populated continent. The population is growing rapidly as **birth rates** are high. Families are often large and about half the population is young and aged below 15 years. Most Africans live in the countryside but a growing number are now found in the towns and cities.

Africa is the poorest inhabited continent in the world. Many of its people live in extreme poverty with a very low **standard of living** and **quality of life**. Some progress has been made in recent years, however, with signs of economic growth and improved living conditions in several countries. This improvement is largely due to changes in government policies and a better use of resources.

B Old Soweto Market, Lusaka, Zambia

Activities

1   Give the latitude and longitude of each of the following cities. Pages 10 and 11 will help you.
    a Cairo   b Nairobi
    c Cape Town   d Dakar

2   Name the countries that have a border with:
    a Kenya   b South Africa   c Libya

3   Describe Africa's human features using the headings below. Write a few short statements for each one.

> **Africa: human features**
> * Population
> * Problems
> * Industry and resources
> * Signs of improvement

4   Write short descriptions of photographs **A**, **B** and **C**. Pages 110 and 111 will help you.

C Waterfront, Cape Town, South Africa

Africa – human features

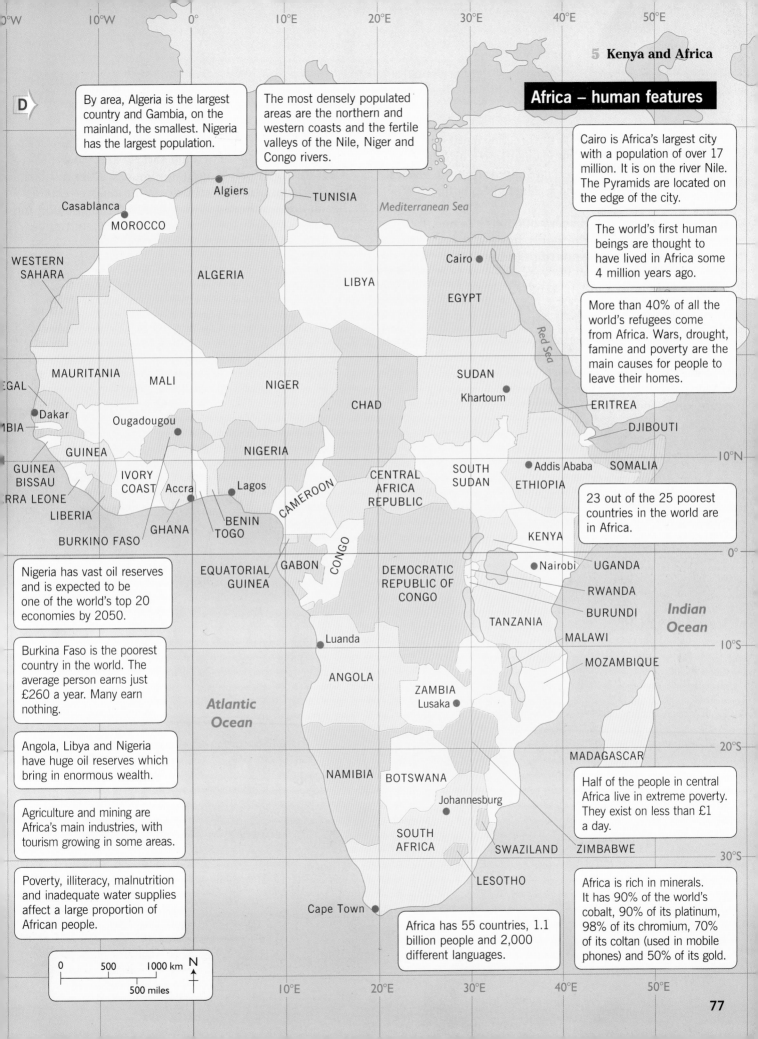

By area, Algeria is the largest country and Gambia, on the mainland, the smallest. Nigeria has the largest population.

The most densely populated areas are the northern and western coasts and the fertile valleys of the Nile, Niger and Congo rivers.

Cairo is Africa's largest city with a population of over 17 million. It is on the river Nile. The Pyramids are located on the edge of the city.

The world's first human beings are thought to have lived in Africa some 4 million years ago.

More than 40% of all the world's refugees come from Africa. Wars, drought, famine and poverty are the main causes for people to leave their homes.

23 out of the 25 poorest countries in the world are in Africa.

Nigeria has vast oil reserves and is expected to be one of the world's top 20 economies by 2050.

Burkina Faso is the poorest country in the world. The average person earns just £260 a year. Many earn nothing.

Angola, Libya and Nigeria have huge oil reserves which bring in enormous wealth.

Agriculture and mining are Africa's main industries, with tourism growing in some areas.

Poverty, illiteracy, malnutrition and inadequate water supplies affect a large proportion of African people.

Half of the people in central Africa live in extreme poverty. They exist on less than £1 a day.

Africa is rich in minerals. It has 90% of the world's cobalt, 90% of its platinum, 98% of its chromium, 70% of its coltan (used in mobile phones) and 50% of its gold.

Africa has 55 countries, 1.1 billion people and 2,000 different languages.

Casablanca
MOROCCO
Algiers
TUNISIA
Mediterranean Sea
WESTERN SAHARA
ALGERIA
LIBYA
Cairo
EGYPT
Red Sea
MAURITANIA
MALI
NIGER
CHAD
SUDAN
Khartoum
ERITREA
DJIBOUTI
EGAL
Dakar
MBIA
Ougadougou
GUINEA
NIGERIA
SOUTH SUDAN
ETHIOPIA
Addis Ababa
SOMALIA
GUINEA BISSAU
IVORY COAST
Accra
Lagos
CAMEROON
CENTRAL AFRICA REPUBLIC
KENYA
RRA LEONE
LIBERIA
GHANA
BENIN
TOGO
BURKINO FASO
EQUATORIAL GUINEA
GABON
CONGO
DEMOCRATIC REPUBLIC OF CONGO
Nairobi
UGANDA
RWANDA
BURUNDI
TANZANIA
Indian Ocean
Luanda
MALAWI
MOZAMBIQUE
ANGOLA
ZAMBIA
Lusaka
Atlantic Ocean
MADAGASCAR
NAMIBIA
BOTSWANA
Johannesburg
SOUTH AFRICA
SWAZILAND
ZIMBABWE
LESOTHO
Cape Town

0   500   1000 km
500 miles
N

What are Kenya's main features?

'Jambo' means hello and is the greeting always given by Kenyans. It is usually accompanied with a broad smile and an offer of help. Indeed most Kenyans seem always to be cheerful, relaxed and willing to help others.

Yet life in Kenya can be very difficult. It is one of the poorest countries in the world, has few services, little industry and a low **standard of living**. It is called a **developing country** and is very different from the UK which is rich, has a high standard of living and is an example of a **developed country**.

Despite being poor in terms of wealth, Kenya is rich in scenery and wildlife and has become a popular tourist destination. Inland there are high mountains, volcanoes, lakes and grassy plains. On the coast there are long sandy beaches, coral reefs and tropical forest.

Most tourists are attracted by the wildlife. They go on organised tours called **safaris** and view the animals from open-topped mini-buses. After a safari, many visitors travel to the coast for a more relaxing beach holiday. Because Kenya lies on the Equator, the climate is usually hot and sunny throughout the year.

Kenya also has a varied population which is made up of people from several different groups or tribes.

Each group has its own **ethnic** background with its own language, religion and way of life. Two of the best known tribes are the Maasai and the Kikuyu.

Kenya's two major cities are Nairobi, the capital, and Mombasa on the coast.

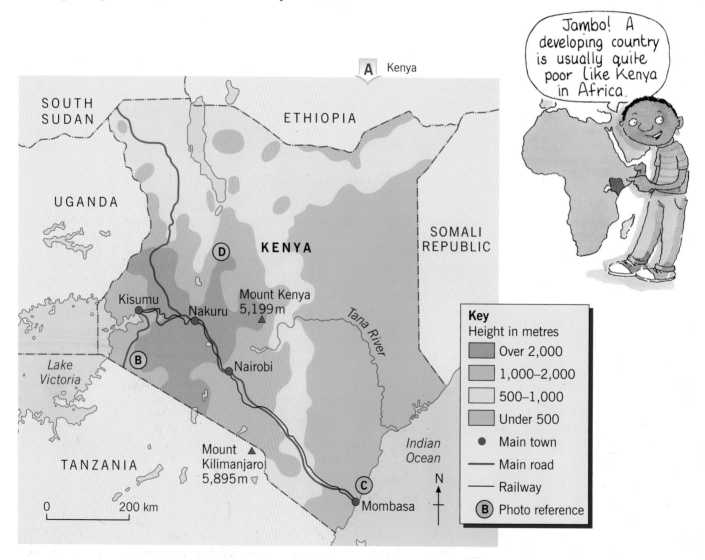

A Kenya

Jambo! A developing country is usually quite poor like Kenya in Africa.

Key

Height in metres

- Over 2,000
- 1,000–2,000
- 500–1,000
- Under 500
- ● Main town
- ── Main road
- ── Railway
- Ⓑ Photo reference

Activities

1 Give the meaning of the following terms. The Glossary at the back of the book may help you.

   a developing country   b developed country

   c standard of living   d ethnic

2 Look at map **A** opposite.

   a Name the five countries that border Kenya.

   b Name the four main towns in order of height above sea level. Give the highest first.

   c How long is Kenya's coastline?

   d How far is it by rail from Kisumu to Mombasa?

   e What is the furthest distance from north to south?

3 Copy and complete quizword **E**, using the following clues.

   (a) Kenya's highest mountain

   (b) A line of latitude across Kenya

   (c) An ocean off Kenya's east coast

   (d) A holiday where wild animals are viewed

   (e) Kenya's main port

   (f) A traditional Kenyan greeting

   (g) A Kenyan tribe

   Make up a clue for downword (h).

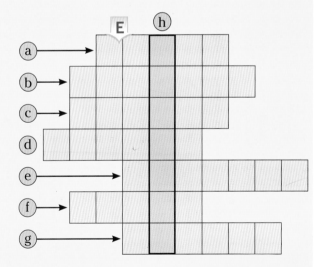

4 Imagine you are on holiday in Kenya. Use the information on this page to help you write a postcard home which describes the country.

Summary

Kenya is a developing country. Although it is not rich, it does have a wealth of spectacular scenery and wildlife.

What are Kenya's main physical features?

Most visitors to Kenya fly in to Nairobi, the country's largest city. Nairobi is in a region called the Central Highlands which millions of years ago was an area of considerable **volcanic activity**.

During that time huge cracks developed in the earth's surface. Two of the cracks ran from north to south through Kenya and as the land between the cracks collapsed, a huge **rift valley** was formed. Earth movements and volcanic eruptions are now quite rare but the cone-shaped remains of volcanoes can be seen throughout the area. Mount Kenya at 5,899 metres is the highest and best known of these.

The scenery and vegetation change dramatically towards the coast. East of the Central Highlands the land is more gently sloping and there are huge areas of grassland. A band of lush, green, tropical **rainforest** lies along the coast where the climate is very hot and wet.

A | Mount Kenya

B

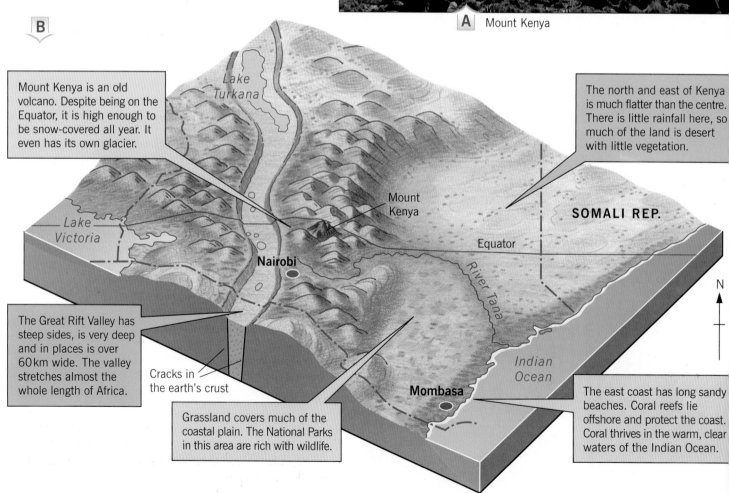

Mount Kenya is an old volcano. Despite being on the Equator, it is high enough to be snow-covered all year. It even has its own glacier.

The north and east of Kenya is much flatter than the centre. There is little rainfall here, so much of the land is desert with little vegetation.

The Great Rift Valley has steep sides, is very deep and in places is over 60km wide. The valley stretches almost the whole length of Africa.

Cracks in the earth's crust

Grassland covers much of the coastal plain. The National Parks in this area are rich with wildlife.

The east coast has long sandy beaches. Coral reefs lie offshore and protect the coast. Coral thrives in the warm, clear waters of the Indian Ocean.

Lake Turkana

Lake Victoria

Mount Kenya

Nairobi

Equator

SOMALI REP.

River Tana

Indian Ocean

Mombasa

N

Kenya is on the Equator which means that average temperatures are high throughout the year. There are no winters or summers as there are in Britain. In terms of temperature, one day is very similar to the next. Seasons are defined by rainfall amounts. The rainy season is usually from April to May.

A typical day in Nairobi begins with bright, clear weather. By the afternoon clouds will have built up and there may be a shower of rain. In the evening it becomes clear and dry again, though rather cool.

Not all of Kenya is like this. The north is hot and dry and rarely has rain. Further south and along the coast there is plenty of sunshine, high temperatures throughout the year, and more rainfall. Afternoon sea-breezes cool the air on the coast but evenings are warm and very pleasant.

With attractive scenery and a climate like this, it is not surprising that Kenya's coastline has become increasingly popular with tourists.

C Climate graphs

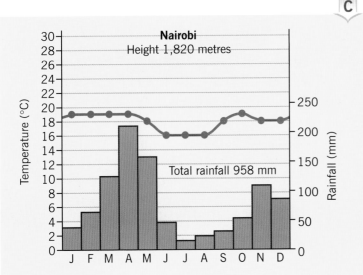

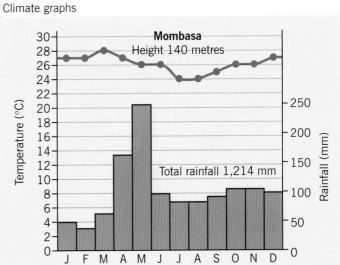

Activities

1 a What is the Great Rift Valley?
   b With the help of the diagrams below, explain how the rift valley was formed.

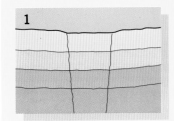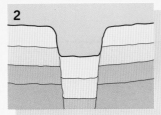

2 Imagine that you have just taken photo **A** of Mount Kenya. Write a caption for the photo to both describe **and** explain the mountain's main features. Include the following:

- steep
- rocky
- volcano
- Equator
- cold
- cone-shaped
- snow-covered
- 5,899 m high

3 Look at the climate graph for Nairobi above.
   a Which three months are the wettest?
   b How much rainfall is there in July?
   c What temperature would you expect in July?
   d Why do you think Nairobi is cooler than Mombasa?

4 a Describe the attractions for tourists to Kenya's coastline. Use the following headings:
      - General weather
      - July temperature
      - July rainfall
      - Coastal landforms
      - Vegetation cover
      - Sea conditions.
   b When is a bad time to take a holiday in the area? Give reasons for your answer.

Summary

Many of Kenya's most attractive landforms can be seen either in the Central Highlands or along the coast. Most of Kenya is hot all year but a lack of rain can be a problem.

Why is Kenya's population unevenly spread?

The population of Kenya, as in most other countries, is not spread out evenly. Some places are very crowded while others have very few people living there. This is mainly due to:

- **migration** – where the people of Kenya originally came from
- **physical conditions** – differences in climate and relief in Kenya.

The movement of people into Kenya

Most present-day Kenyans are descended from African tribes who arrived in the country from four main directions (map **A**). Most came from the south, making this the most densely populated area.

While individual tribes still remain today, their way of life has been changed through marriages and contact with people from other ethnic groups.

Differences in physical conditions

Although Kenya is not a very large country, the physical conditions vary considerably from place to place. Map **B** shows how the distribution of population is affected by differences in rainfall, temperature, water supply, relief, soil and vegetation.

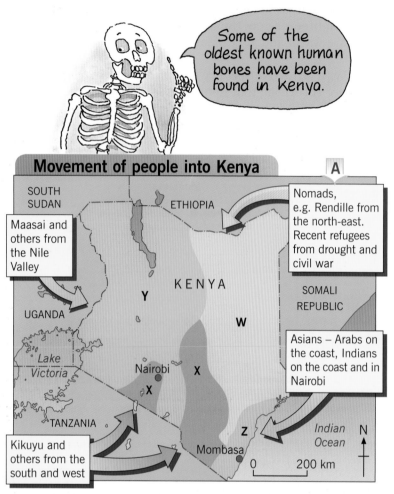

Some of the oldest known human bones have been found in Kenya.

Movement of people into Kenya **A**

SOUTH SUDAN

ETHIOPIA

Nomads, e.g. Rendille from the north-east. Recent refugees from drought and civil war

Maasai and others from the Nile Valley

K E N Y A

Y

W

SOMALI REPUBLIC

UGANDA

Lake Victoria

Nairobi

X

X

Asians – Arabs on the coast, Indians on the coast and in Nairobi

TANZANIA

Z

Indian Ocean

N

Mombasa

0   200 km

Kikuyu and others from the south and west

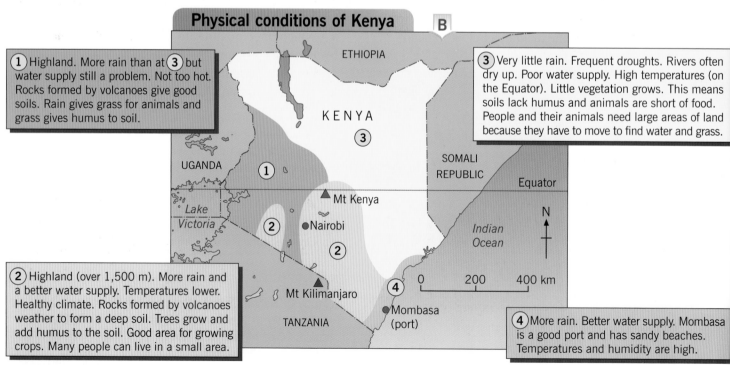

Physical conditions of Kenya **B**

① Highland. More rain than at ③ but water supply still a problem. Not too hot. Rocks formed by volcanoes give good soils. Rain gives grass for animals and grass gives humus to soil.

ETHIOPIA

③ Very little rain. Frequent droughts. Rivers often dry up. Poor water supply. High temperatures (on the Equator). Little vegetation grows. This means soils lack humus and animals are short of food. People and their animals need large areas of land because they have to move to find water and grass.

K E N Y A

③

UGANDA

①

SOMALI REPUBLIC

Equator

Lake Victoria

② Mt Kenya

②

Nairobi

②

Indian Ocean

N

② Highland (over 1,500 m). More rain and a better water supply. Temperatures lower. Healthy climate. Rocks formed by volcanoes weather to form a deep soil. Trees grow and add humus to the soil. Good area for growing crops. Many people can live in a small area.

Mt Kilimanjaro

④

0   200   400 km

TANZANIA

Mombasa (port)

④ More rain. Better water supply. Mombasa is a good port and has sandy beaches. Temperatures and humidity are high.

Present-day movements of population

In all developing countries there is a large movement of people from the countryside to the towns. This is called **rural-to-urban migration**. In Kenya it is mainly the Kikuyu who move. Their traditional home is in area ② on map **B**.

When driving through rural Kikuyu countryside it is hard to see why they want to move. It is one of the few parts of Kenya with roads, it has the best farmland and water supply in the country and the environment appears clean and pleasant. To those living there, especially those at school or just starting a family, it is less attractive.

Photo **C** gives some of the reasons why many Kikuyu want to move to Nairobi, the capital of Kenya.

Nairobi has big, modern buildings which include hospitals, shops, cinemas and a university.

The Kikuyu have lived in villages and small towns for a long time. The change to city life should be easy.

Kenya has one of the highest birth rates in the world. The average family size is 4.8 people. There are too many of us to find jobs on the farms and in the shambas.

It is not far to Nairobi so we can get work there and still visit our village.

We are farmers but many of us do not own any land and if we do, the plots are very small.

We have learned some skills at school but we cannot use them in our local village.

Activities

1   Read the following statements about Kenya's population. Write out the four statements that are correct.
    - Kenya's population is spread evenly.
    - Kenya's population is not spread evenly.
    - The Maasai came from the Nile Valley and live in the south-west.
    - The Maasai came from the south-west and live in the Nile Valley.
    - The Kikuyu live in the north.
    - The Kikuyu came from the south and west and live on higher land.
    - Arabs and Indians live on the coast near Mombasa.

2   Make a copy of diagram **D** and complete it to show why the south of Kenya is more crowded than the north. For each box choose the correct word from the two given in brackets. 'Relief' has been done for you.

3   Imagine that you live in a small Kikuyu village and are about to migrate to Nairobi.
    a   Give at least three reasons for leaving your village.
    b   Give at least three advantages of living in Nairobi.
    Diagram **C** will help you to answer this question.

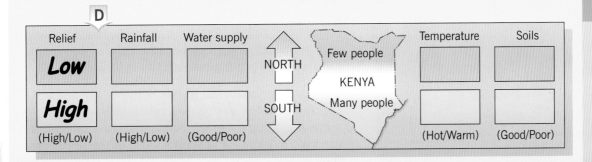

Relief	Rainfall	Water supply			Temperature	Soils
Low			NORTH	Few people		
				KENYA		
High			SOUTH	Many people		
(High/Low)	(High/Low)	(Good/Poor)			(Hot/Warm)	(Good/Poor)

Summary

The distribution of population in Kenya is mainly affected by physical factors such as climate, water supply, relief and soil. Most people live in or near to the capital city of Nairobi.

What is it like living in Nairobi?

Like most cities in developing countries there are two sides to Nairobi. One side is seen by overseas visitors and the relatively few wealthy Kenyans. Photo **A** shows the most important building in Nairobi. It is the Kenyatta International Conference Centre, named after Jomo Kenyatta, the first president of Kenya. Around it, in central Nairobi, are tall, modern buildings and wide, tree-lined streets.

The other side of Nairobi is the one seen by most of the Kenyans who migrate from the surrounding rural areas. Many migrants move to live with family and friends already living in Nairobi. They share houses, food and even jobs. In time the newcomers may be able to construct their own homes in one of several **shanty settlements** found on the edge of the city (photo **B**).

Living in Kibera

Kibera is one shanty settlement. It is 6 kilometres and a 10 pence bus ride from the city centre. However, most inhabitants who wish to make that journey have to walk because they cannot afford the fare.

Houses in Kibera are built close together. Sometimes it is hard to squeeze between them. The walls are usually made from mud and the roofs from corrugated iron (photos **C** and **E**). Inside there is often only one room. Very few homes have water, electricity or sewage. One resident, who is considered rich by Kibera standards, has a tap and can sell water to his neighbours. He also has a toilet but this is only emptied four times a year.

Sewage runs down the tracks between the houses (photo **C**). In the wet season rain mixes with the sewage making the tracks unusable, so small children are kept indoors for several weeks. In the dry season the tracks become very dusty. Photo **D** shows an open drain along a main sidetrack being cleared. What might happen to the sewage the next time it rains?

Kenya has one of the highest birth rates in the world. In Kibera it is not unusual for families to have more than ten children. Very few children can read or write as there is only one small school. Many suffer from a poor diet or malaria. Others catch diseases by drinking dirty water or playing in sewage (photo **C**).

People have to find their own way to earn money. Some have small stalls from which they sell food. Others, who have learnt a skill, may turn their houses into shops (photos **B** and **E**) or collect waste material and recycle it in small workshops.

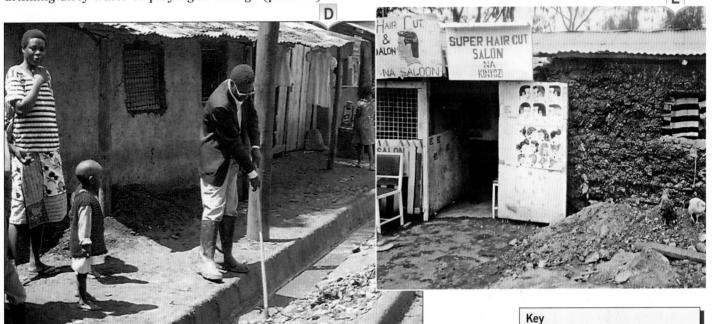

Activities

1   Give three differences between photos **A** and **B**.

2   Map **F** shows the location of shanty settlements in Nairobi.
   a   How far is Kibera from the city centre?
   b   Which direction is Kibera from the city centre?
   c   Write out the following sentence using the correct word from the pair in brackets.

   Shanty settlements are areas of (good/poor) quality housing found at the (edge/middle) of the city on (good/poor) quality building land.

3   Using photo **C**, list four problems of life in a shanty settlement.

4   List four ways that people living in a shanty settlement can earn money.

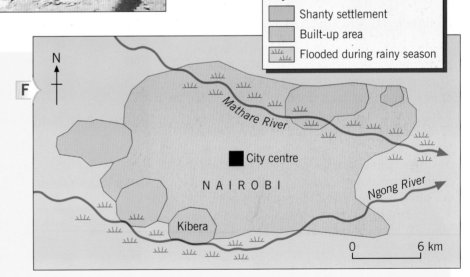

Key
Shanty settlement
Built-up area
Flooded during rainy season

5   a   List four good points about living in Nairobi.
   b   Why is Nairobi described as having 'two sides'?

Summary

Cities in developing countries have two sides. Well-off people live and work in good conditions near to the city centre. Poor people often live and work in less pleasant shanty settlements a long way from the city centre.

What is the Maasai way of life?

The Maasai in Kenya

One ethnic group living in Kenya are the Maasai (see map **A** on page 82). They are **pastoralists** with herds of cattle and goats. Some are **nomadic** and have to move about to find water and grass for their animals. The Maasai depend on these animals for their daily food. It is cattle, not money, which means wealth to the Maasai. Indeed, the usual Maasai greeting is 'I hope your cows are well'.

The land where the Maasai live is fairly flat and covered with grass which depends upon the rain. The 'long rains' come between April and June, the 'short rains' in October and November. In the dry months the grass withers under the hot sun. If the rains do not come then the Maasai have to move to look for grass for their animals.

Houses

Most Maasai live in an **enkang** which is a small village made up of 20 to 50 huts (photo **A**), in which 10 to 20 families live. It is surrounded by a thick thorn hedge to keep out dangerous animals such as lions, leopards and hyenas. Tiny passages allow people and, in the evening, cattle to pass through. These passages are blocked up at night. The huts are built in a circle around an open central area. They barely reach the height of an adult Maasai (photo **B** and plan **D**) and are built from local materials.

The frame is made from wooden poles. Mud, from nearby rivers, and cow dung are used for the walls. Grass from the surrounding area is used for the roof. The hut is entered by a narrow tunnel. Apart from an opening the size of a brick, there are no windows or chimneys. The inside is dark and full of smoke from the fire. It is cool during the day, warm at night and free from flies and mosquitoes. Cowskins are laid on the floor for beds. Water and honey are stored in gourds – a ball-shaped plant with a thick skin.

A

B

C

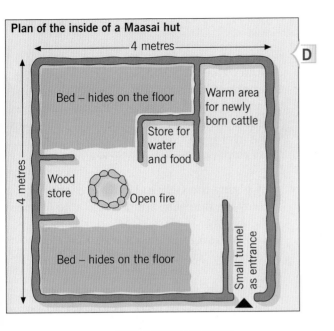

Plan of the inside of a Maasai hut

4 metres

4 metres

D

Bed – hides on the floor

Warm area for newly born cattle

Store for water and food

Wood store

Open fire

Bed – hides on the floor

Small tunnel as entrance

E

Dress

Men wear brightly coloured 'blankets' and women wear lengths of cloth (photo **A**). The women shave their heads and wear beads around their neck, wrists and ankles. Teeth are cleaned with sticks and the men also use sticks to comb their hair. As water is often scarce, the Maasai use animal and vegetable fat to clean themselves and sweet smelling grasses for perfume.

Daily jobs

The women have to collect sticks for the fire and water for cooking. They also make baskets and jewellery. The men spend the whole day guarding their animals. The Maasai regard the ground as sacred and believe it should not be broken. This means no crops can be grown, wells cannot be dug and, often, the dead are not buried but are left to wild animals.

As crops are not grown sometimes animals are exchanged for grain.

Diet

A major part of the Maasai diet is milk mixed with blood from their cows. In times of drought only blood is used.

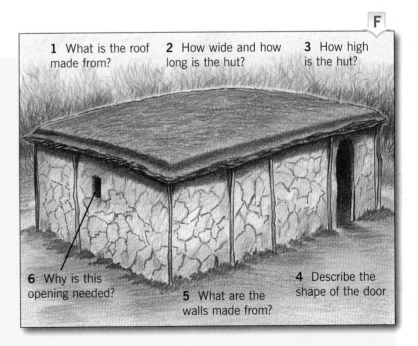

F

1 What is the roof made from?

2 How wide and how long is the hut?

3 How high is the hut?

6 Why is this opening needed?

5 What are the walls made from?

4 Describe the shape of the door

Activities

1 a Give two facts about a developing country. Page 78 and the Glossary will help you.

   b What is meant by an 'ethnic group'?

   c In which continent is Kenya?

2 a Give two reasons why cattle are important to the Maasai.

   b Write a paragraph to describe Maasai farming. Include these words:

   - cattle and goats
   - grass
   - flat land
   - rain
   - nomadic

3 Sketch **F** shows a Maasai hut and several questions. Draw the hut and add labels by answering the questions.

4 How do photo **A** and your drawing of a Maasai hut suggest that the weather is:

   a usually warm   b not very wet?

Summary

Landscape, weather and wealth all affect the family life, housing, clothing and diet of the Maasai.

What is a developing country?

By now you should be aware of many differences between living in Kenya and in a more developed country like the UK, for example. These differences include ethnic groups, dress, housing, jobs, wealth and the quality of life. Kenya is an example of a **developing country**. What is a developing country? How is life in a developing country different from life in a developed country?

In a developed country most, but not all, people earn a lot of money compared with those in a developing country. They live in good houses, have their own cars, televisions, the latest technology and can afford good food and holidays. Compared with Kenya most people in a developed country have a high **standard of living**. Kenya is considered to be 'poor' and developed countries to be 'rich'. Most people see the difference in wealth as the main difference between a developing country and a developed country.

The wealth of a country is given by its **gross national product** (GNP). This is the total amount of money made by a country from its raw materials, its manufactured goods and its services.

Notice that GNP is always given in American dollars (US$). The total amount can then be divided by the total number of people living in that country. This gives the average amount of money available for every person living in the country.

By giving the GNP in US$ it is easy to compare different countries. Table **A** gives the average income (GNP) per person for five developing countries and three developed countries.

Apart from wealth there are many other ways of trying to measure the level of a country's development (table **B**).

A

	Country	GNP (US$ per person)
Developing countries	Bangladesh	747
	Brazil	11,340
	Egypt	3,187
	Kenya	862
	Peru	6,573
Developed countries	Japan	46,720
	UK	38,514
	USA	49,965

B

Jobs		Primary activities give most jobs in a developing country. A developed country has fewer primary activities and more secondary and service jobs.
Trade		A developing country usually has to sell raw materials at a low price and has to buy manufactured goods from developed countries at a high price.
Population		A developing country has a higher **birth** and **death rate**, more young children dying (high **infant mortality**), adults dying at a younger age (short **life expectancy**) and a faster population increase than a developed country.
Health		A developing country has less money to spend on training doctors and nurses and in providing hospitals and medicines.
Education		The large number of children and lack of money for schools in a developing country mean fewer people can read and write than in a developed country (low **literacy rate**).

The 'richest' and 'poorest' countries in the world

What happens when the different methods used in table **B** to measure the level of development of a country are put together? The result is that some parts of the world are made up of countries that are mainly 'rich' whilst other parts are made up of countries that are mainly 'poor'. Map **C** shows this pattern.

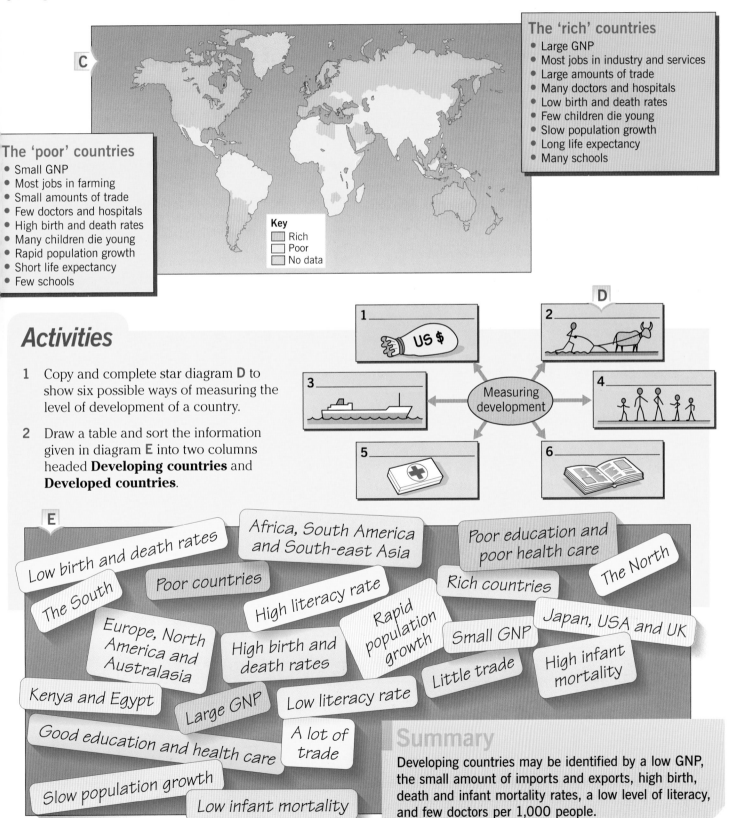

C

The 'rich' countries
- Large GNP
- Most jobs in industry and services
- Large amounts of trade
- Many doctors and hospitals
- Low birth and death rates
- Few children die young
- Slow population growth
- Long life expectancy
- Many schools

The 'poor' countries
- Small GNP
- Most jobs in farming
- Small amounts of trade
- Few doctors and hospitals
- High birth and death rates
- Many children die young
- Rapid population growth
- Short life expectancy
- Few schools

Key
- Rich
- Poor
- No data

D

1. US $
2.
3.
4.
5.
6.

Measuring development

Activities

1 Copy and complete star diagram **D** to show six possible ways of measuring the level of development of a country.

2 Draw a table and sort the information given in diagram **E** into two columns headed **Developing countries** and **Developed countries**.

E

- Low birth and death rates
- Africa, South America and South-east Asia
- Poor education and poor health care
- The South
- Poor countries
- Rich countries
- The North
- High literacy rate
- Europe, North America and Australasia
- High birth and death rates
- Rapid population growth
- Small GNP
- Japan, USA and UK
- Kenya and Egypt
- Large GNP
- Low literacy rate
- Little trade
- High infant mortality
- Good education and health care
- A lot of trade
- Slow population growth
- Low infant mortality

Summary

Developing countries may be identified by a low GNP, the small amount of imports and exports, high birth, death and infant mortality rates, a low level of literacy, and few doctors per 1,000 people.

We have already seen that all countries are different. Some are rich and have high standards of living whilst others are poor and have low standards of living. Countries that differ in this way are said to be at different stages of development.

In this unit you have learned much about Kenya. You will certainly know by now that Kenya is very different from the UK and also, certainly in terms of wealth, that it is very poor. Indeed Kenya is an almost perfect example of a **developing country**.

As we have seen, however, measuring development can be difficult. Development, after all, is about quality of life. Some countries may be economically poor, but their people can still be cheerful, relaxed and generally happy with their lives.

In this enquiry you work for a department of the British Government responsible for overseas development and have been asked to make a report on Kenya's level of development. The report will be in four parts, as shown in drawing **A**. Pages 88–89 will be helpful to you.

British Overseas Aid

A

The Kenya Report

1 How developed is Kenya compared with the UK?

2 How developed is Kenya compared with neighbouring countries in Africa?

3 How developed is Kenya in terms of social and cultural measures of development?

4 In which areas of development (as shown in table **B**) is Kenya in greatest need of improvement? What might be done to help the country make progress?

In each case suggest reasons for your answer. You should consider economic, social and cultural factors.

- **Economic factors** are about the wealth of a country.

- **Social factors** are concerned with standards of living and quality of life.

- **Cultural factors** are about traditions and the way of life.

How developed is Kenya?

1 a Make a copy of table **B** below.

   b Using information from diagram **C**, complete your table to show the rank order of the seven countries for each measure of development.
   The most developed will score 1 and the least developed will score 7.

   c Add the scores together for each country, and complete the Total column.

2 Write out the countries from your completed table as a 'League table of development'. The country with the lowest score will be most developed and should be at the top of the league.

3 Complete the report using your answer to activity **1** and information in diagrams **C** and **D**.

B	Country	Wealth (highest first)	Trade (highest first)	Life expectancy (highest first)	People/doctor (lowest first)	Literacy rate (highest first)	TOTAL
	UK						
	Ethiopia						
	Kenya						
	Somali Rep.						
	Sudan						
	Tanzania						
	Uganda						

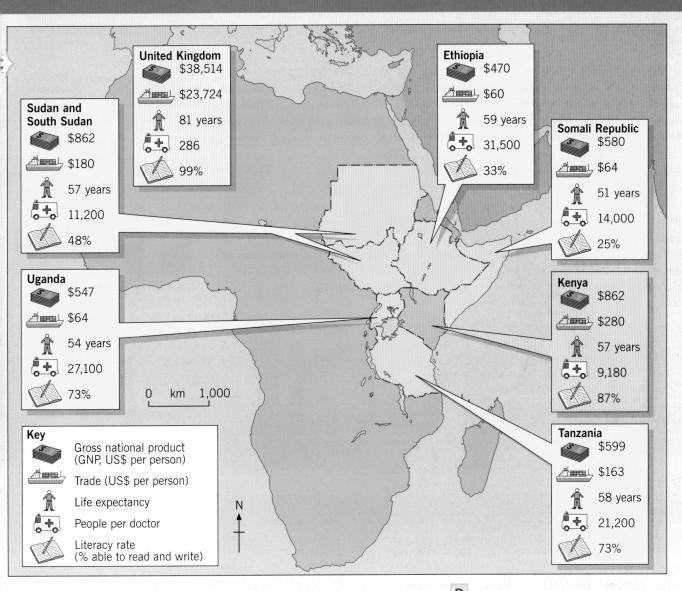

United Kingdom
- $38,514
- $23,724
- 81 years
- 286
- 99%

Ethiopia
- $470
- $60
- 59 years
- 31,500
- 33%

Sudan and South Sudan
- $862
- $180
- 57 years
- 11,200
- 48%

Somali Republic
- $580
- $64
- 51 years
- 14,000
- 25%

Uganda
- $547
- $64
- 54 years
- 27,100
- 73%

Kenya
- $862
- $280
- 57 years
- 9,180
- 87%

Tanzania
- $599
- $163
- 58 years
- 21,200
- 73%

0   km   1,000

Key
- Gross national product (GNP, US$ per person)
- Trade (US$ per person)
- Life expectancy
- People per doctor
- Literacy rate (% able to read and write)

N

D

Our country is making progress but most of us are still very poor.

We have a wealth of beautiful scenery and exciting wildlife.

Family values are important to us and we are always willing to help others.

We still need a clean, reliable water supply and more food.

Kenyan children always seem to have a smile on their faces and be full of fun.

Our traditional way of life is interesting and colourful.

We have some of the best athletes in the world.

People in Kenya are well known for being cheerful and friendly.

Maps show what things look like from above. They are very useful because they give information and show where places are. There are many different types of map. These include street maps, road maps, **atlas** maps and **Ordnance Survey** (OS) maps.

A **plan** is a type of map. Plans give detailed information about small areas. Places like schools, shopping centres, parks and leisure centres are shown on plans.

This section is about **direction**. The best way to show direction is to use the **points of the compass**. There are four main points. These are north, east, south and west. You can remember their order by saying 'Never Eat Shredded Wheat'.

Between these four main points there are four other points. These are north-east, south-east, south-west and north-west.

Most maps have a sign to show the **north** direction. If there is no sign the top edge of the map should be **north**.

A

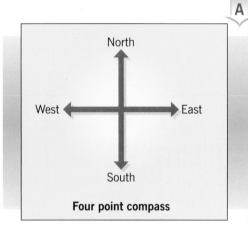

Four point compass

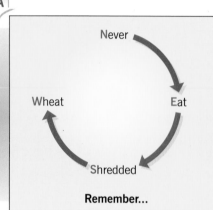

Remember...

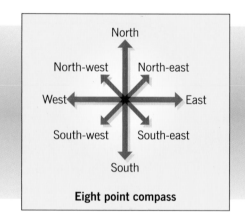

Eight point compass

To give direction for a place you have to say which way you need to go to get there. The direction is the point of the compass towards which you have to go. Diagrams **B**, **C** and **D** show you how to give a direction.

B

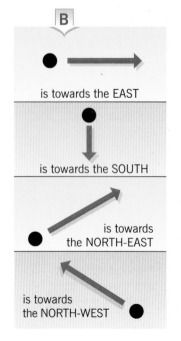

is towards the EAST

is towards the SOUTH

is towards the NORTH-EAST

is towards the NORTH-WEST

C

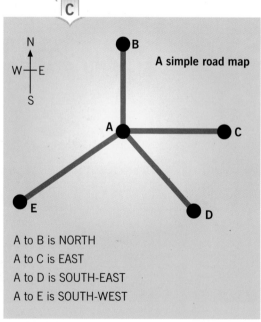

A simple road map

A to B is NORTH
A to C is EAST
A to D is SOUTH-EAST
A to E is SOUTH-WEST

D

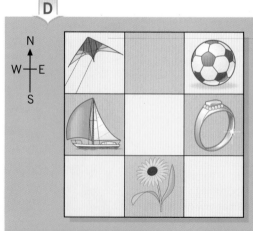

The kite is NORTH of the boat.
The ring is SOUTH of the ball.
The boat is WEST of the ring.
The flower is SOUTH-EAST of the boat.
The ring is NORTH-EAST of the flower.

Activities

1 Draw the compass in diagram **E** and label the unmarked points.

2 Copy these drawings and complete the sentences below them. The first one has been done for you.

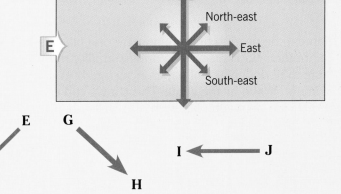

B is north of **A**   **D** is . . . of **C**   **F** is . . . of **E**   **H** is . . . of **G**   **I** is . . . of **J**

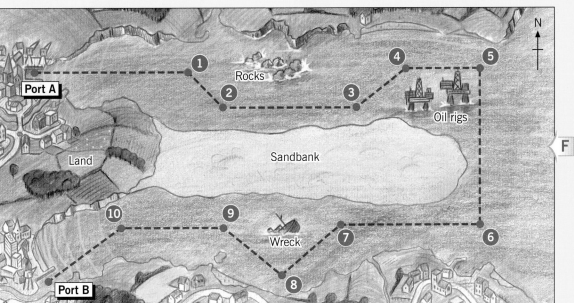

3 Study map **F** and give the following directions:
  a from Port A to the rocks
  b from the wreck to the oil rigs
  c from the oil rigs to the rocks
  d from the wreck to Port A
  e from the rocks to the wreck.

4 a A ship has landed its cargo at Port A. It must go to Port B to reload. The course the ship must follow is shown by the dotted line on the map. Give the Captain compass directions to follow between each numbered point.
    Start like this: *Leave Port A. Go east to point 1. Go south-east . . .*

  b Imagine that the sandbank has been cleared to make ship movement easier. Work out the best course from Port B to Port A. Give compass directions to follow that course.

5 You will need to use the Ordnance Survey map of the Cambridge area for this activity. It is on the inside back cover.

Look at the villages near the bottom of the map. Give the following directions:
  a from Foxton to Whittlesford
  b from Foxton to Newton
  c from Great Shelford to Whittlesford
  d from Great Shelford to Haslingfield
  e from Haslingfield to Harston.

Summary

Maps are a good way of giving information and showing where places are. Direction can be described by using the points of the compass.

How can we measure distance?

A map can be used to find out how far one place is from another. Maps have to be drawn smaller than real life to fit on a piece of paper. How much smaller they are is shown by the **scale**. This shows you the **real** distance between places. In diagram **A** the scale line shows that 1 cm on the map is the same as 1 km on the ground. Every map should have a **scale line**.

Straight line distances are easy to work out. Diagram **A** shows how to measure the straight line, or shortest, distance between the school and the bridge.

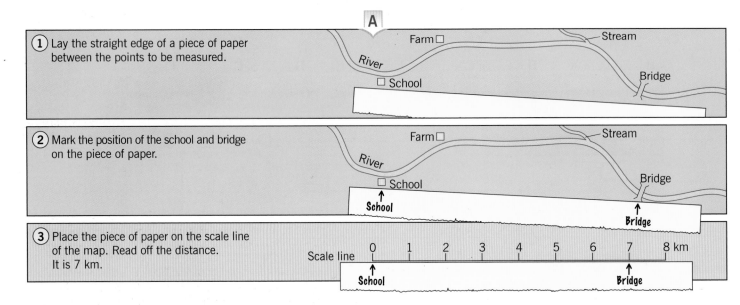

A

① Lay the straight edge of a piece of paper between the points to be measured.

② Mark the position of the school and bridge on the piece of paper.

③ Place the piece of paper on the scale line of the map. Read off the distance. It is 7 km.

The same method can be used to work out distances that are not straight lines. To measure these, divide the route into a number of sections and measure each one.

This can be done by using a piece of paper and turning it at each bend. Diagram **B** shows how to measure the distance from the school to the bridge, following the river.

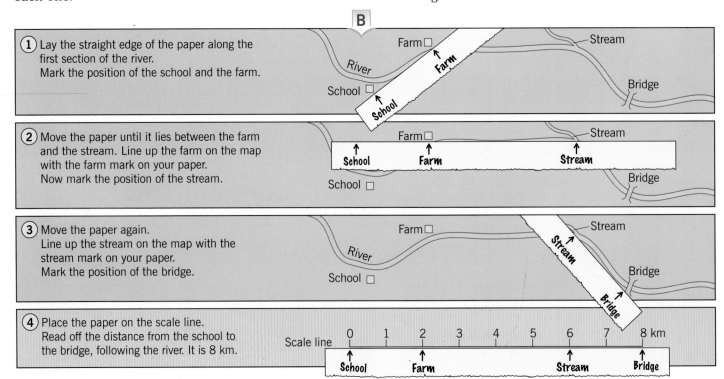

B

① Lay the straight edge of the paper along the first section of the river. Mark the position of the school and the farm.

② Move the paper until it lies between the farm and the stream. Line up the farm on the map with the farm mark on your paper. Now mark the position of the stream.

③ Move the paper again. Line up the stream on the map with the stream mark on your paper. Mark the position of the bridge.

④ Place the paper on the scale line. Read off the distance from the school to the bridge, following the river. It is 8 km.

Activities

1 Use the scale line from map **C** to give the lengths of these lines. Answer like this:

Line a is ... metres (m) in length.

a _____

b _____

c _____

d _____

2 Use map C and the scale line to give the straight line distance between the places below. Choose your answers from the following:

| 40 m | 80 m | 100 m | 120 m |

a Kate's house and the school
b Joanne's house and the post office
c Tim's house and the post office
d John's house and the garage

3 a Give the distance Joanne has to travel to school if she calls on Kate on the way.
   b Give the distance John has to travel to school if he calls at the shop and post office first.

4 What is the distance around the duck pond if you walk on the footpath? Give your answer in metres (m).

5 You have been given a map and some instructions to help you find some hidden treasure. Follow the instructions to find out where it is.

Check the exact spot by sorting out the jumbled words in the treasure chest and choosing the correct answer.

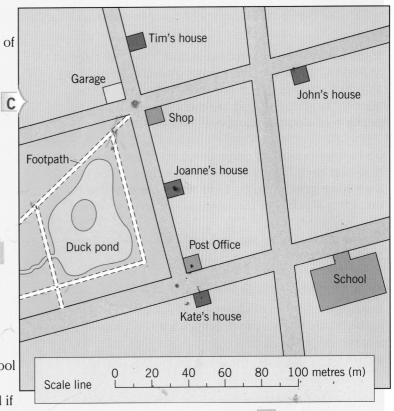

Summary

Distances on a map can be measured using the scale line. The scale line gives the real distance between places on the map.

What are grid references?

Maps can be quite complicated and it may be difficult to find things on them. To make places easier to find, a grid of squares may be drawn on the map. If the lines making up the grid are numbered, the exact position of a square can be given.

On Ordnance Survey maps these lines are shown in blue and each has its own special number. The blue lines form **grid squares**. **Grid references** are the numbers which give the position of a grid square.

On these two pages you will learn about **four figure grid references**.

To *give* a grid reference is simple.

Look at the grid in diagram **A** and follow these instructions to give the reference for the yellow square.

- Give the number of the line on the *left* of the yellow square – it is **04**.
- Give the number of the line at the *bottom* of the yellow square – it is **12**.
- Put the numbers together and you have a four figure grid reference. It is **0412**.

In the same way, the Picnic Square has a reference of 0313 and the School Square is 0512.

What are the grid references for the Bridge Square and also the Tent Square?

To *find* a grid reference is also easy.

Look at the grid in diagram **B** and follow these instructions to find grid square **4237**.

- Go along the top of the grid until you come to **42**. That line will be on the *left* of your grid square.
- Go up the side of the grid until you come to **37**. That line will be at the *bottom* of your square.
- Now follow those two lines until they meet. Your square will be above and to the right of that point. There is a house in it.

What is in squares 4136 and 4037?

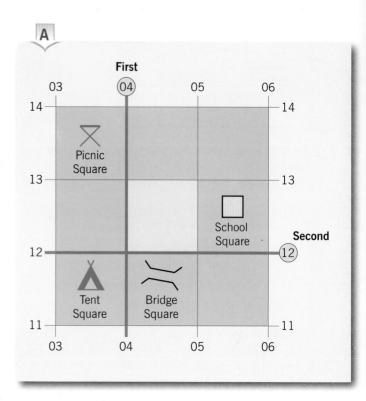

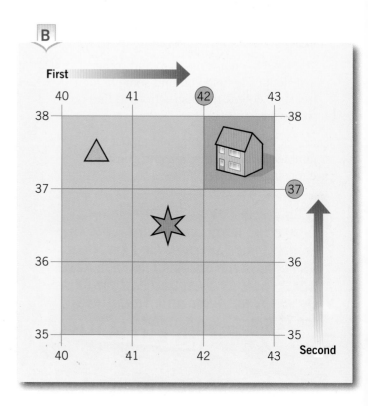

Activities

Look at map **C** of the British Isles. It shows some of the main towns, mountain areas and the three longest rivers. Use the map to answer the questions below.

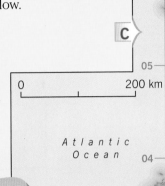

✔ Remember

✔ The line on the left comes first.

✔ The line at the bottom comes second.

It may help you to remember if you say '**Along** the corridor and **up** the stairs'.

1 Name the towns in each of the grid squares given below. Choose your answers from this list:

Belfast   Manchester

Glasgow   Bristol

a 0202 b 0104
c 0200 d 0003.

2 Name the mountain areas in each of the following grid squares:
a 0104 b 0103 c 0202.

3 a Which rivers flow through grid square 0201?
b Which river reaches the sea in grid square 0201?

4 Give the grid references for these places:
a Dublin b Newcastle upon Tyne
c London d The Irish Sea.

5 Give the grid reference for the place where you live.

6 Look at the Ordnance Survey map on the inside back cover. Name the farms in each of the following grid squares (the symbol for farm is Fm).
a 4149 b 4156 c 4456
d 4650 e 4257.

Summary

Grid references can be used to help describe the location of a place on a map.

How do we use six figure grid references?

Grid references are very useful in helping us to find places on maps. A four figure reference on an Ordnance Survey map equals an area on the ground of one square kilometre. This is quite a large area. To be more accurate we need to use a **six figure grid reference**. This pinpoints a place exactly to within 100 metres.

Look at the grid in diagram **A**. The six figure grid reference for the windmill is 045128. Follow these instructions and look at diagrams **B** and **C** to see how that reference is worked out.

- Give the number of the line on the *left* of the yellow square – it is **04**.

- In your head divide the square into tenths as shown in the grid in diagram **B**. Follow arrow **A** across the square. The windmill is about halfway across from the left. That puts it on the five-tenths line. Write down **5** after your number 04.

- You now have the first half of your six figure reference – **045**.

✔ Remember
- ✔ The numbers along the **bottom** come first.
- ✔ The numbers on the **left** come second.
- ✔ There must always be six figures.

- Now give the number of the line at the *bottom* of the yellow square – it is **12**.

- In your head divide the square into tenths as shown in the grid in diagram **C**. Follow arrow **B**. The windmill is over halfway up from the bottom. That puts it on the eight-tenths line. Write down **8** after your number 12.

- You now have the second half of your six figure reference – it is **128**.

- Put the two halves together and you have **045128**.

A
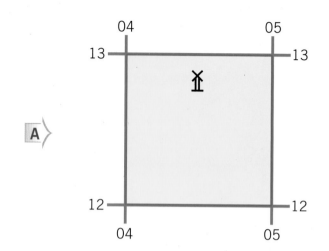

B
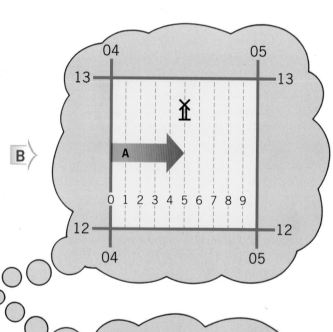

C
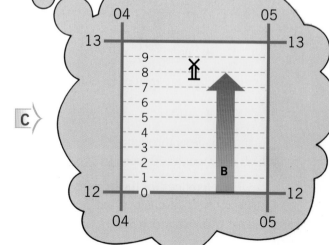

Activities

Look at map **D**. The 'tenths' lines have been added to help you with activities **1**, **2** and **3**. Check your references ...

- The village of Eldon is in grid square 1623.
- The Mill is at reference 166256.
- Dingle Farm is at reference 170238.

1 Copy and complete the sentences below. Use the correct answer from the brackets.
   a At 168245 there is a (church, post office, farm).
   b At 165257 there is a (telephone, school, bridge).
   c At 175233 there is a (farm, lake, level crossing).
   d At 177244 there is a (station, wood, roundabout).

2 Give the six figure grid reference for each of the following:
   a Eldon post office
   b Causey railway station
   c Padley school
   d Burr Wood picnic site.

3 a Follow these directions for a pleasant walk:

   > Start at 170238. Walk down to 173237. Turn left and go to 177244. Go along the road to 171248. Follow the path to 178257. Turn left and finish your walk when the path reaches the road.

   b Name the place where you finished your walk. Give its six figure grid reference.
   c Where would you have stopped for lunch?
   d How many churches did you pass on the way? Give their six figure grid references.

4 You will need to use the Ordnance Survey map of the Cambridge area for this question. It is on the inside back cover.
   a Make a copy of table **E**.
   b Use the map to complete table **E**. The missing symbols, meanings and references are given in diagram **F**.

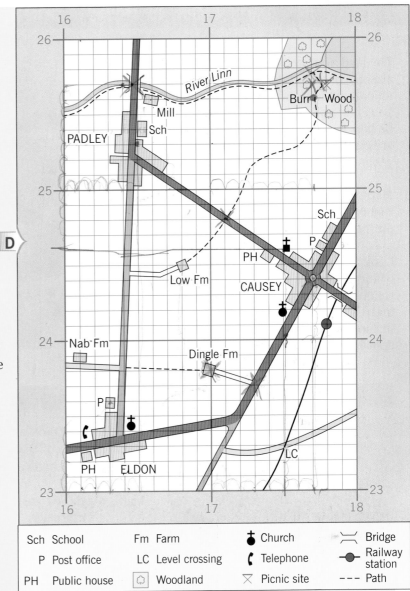

D

Sch	School	Fm	Farm	✝	Church	⌣⌢	Bridge
P	Post office	LC	Level crossing	(	Telephone	●—	Railway station
PH	Public house	⌂	Woodland	✕	Picnic site	- - -	Path

Symbol	Meaning	Six figure grid reference
—●—		465523
◆		
		488505
	Church with tower	
	Camp/caravan site	

E

F
418509
Railway station
Wood
453539
Motorway junction
440534

Summary

Six figure grid references can be used to give the exact position of a place on a map.

How is height shown on a map?

The land around us is seldom flat. There are nearly always differences in height and differences in slope. Sometimes slopes may be gentle and at other times steep. There may be hills, mountains and valleys or areas that are quite level. The word **relief** is used by geographers to describe the shape of the land.

Map makers have to find ways of showing relief and height. How they do this is shown on the next four pages.

Look at sketch **A**. How can height on the island be shown on a flat piece of paper? Height is usually measured from sea level in metres. This can then be shown on a map in three different ways. These are by using **spot heights**, **layer colouring** and **contours**.

A

Spot heights
B

These give the exact height of a point on the map. They are shown as a black dot and each one has a number next to it. The number gives the height in metres. A **triangulation pillar** is also used to show height. These are drawn as a dot inside a blue triangle on the map.

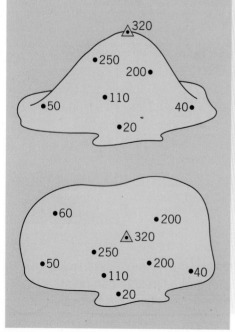

Layer colouring
C

This can also be called **layer shading**. Areas of different heights are shown by bands of different colours. Brown is usually used for high ground, and green for low ground. There always needs to be a key. Layer colouring is used in atlases to show height.

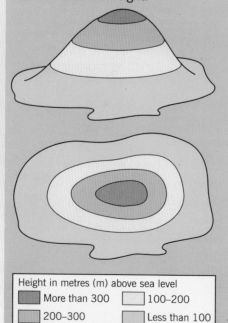

Height in metres (m) above sea level	
▇ More than 300	☐ 100–200
▇ 200–300	☐ Less than 100

Contours
D

Contours are lines drawn on a map. They join places which have the same height. They are usually coloured brown. Most contours have their height marked on them but you may have to trace your finger along the line to find it. Sometimes you will have to go to the contour above or below to get the height. Heights are given in metres.

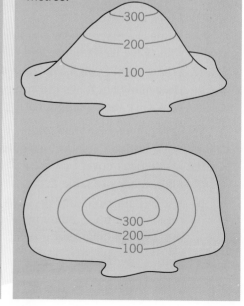

Activities

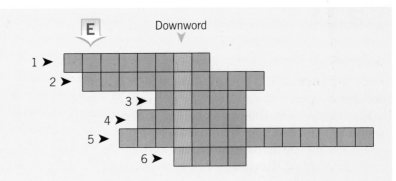

1 **a** Copy out and complete crossword **E** using the clues below.

   b When you have finished, give the meaning of the downword in the orange squares.

   Clues

   1 Lines that join places of the same height.
   2 Height at one place.
   3 This can be gentle or steep.
   4 Measured from sea level.
   5 Colouring to show height.
   6 A level area with no slope.

2 Look at map **F** of England and Wales. The map uses layer colouring to show height. The letters mark land at different heights.

   a Which letters mark lowland areas under 100 metres?

   b Which letters mark land between 100 and 500 metres?

   c Which letters mark land above 500 metres?

3 Use map **F** to answer these questions.

   a The highest mountain in England is Scafell Pike and the highest mountain in Wales is Snowdon.
   What colour are they shaded?

   b The Pennines are an area of high land in the centre of northern England.
   How high are they?

   c The Cotswolds and Chilterns are hills in the south of England.
   What height are they?

   d What height is the area where you live?

4 Look at the Ordnance Survey map on the inside back cover. Give the heights above sea level of the following:

   a the contours in grid squares 4852 and 4450

   b the spot heights in grid squares 4151 and 4754

   c the triangulation pillar in grid square 4051.

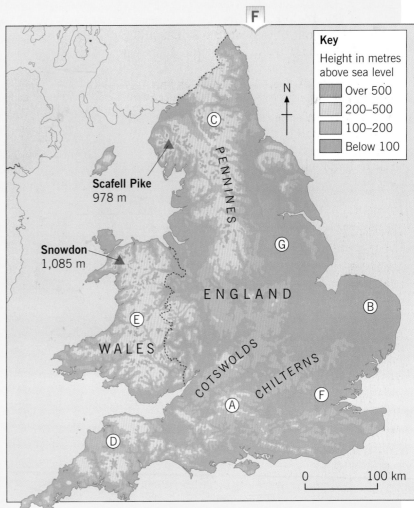

5 On the same Ordnance Survey map, look at Rowley's Hill in grid square 4249.

   a Draw the pattern of contours and the triangulation pillar.

   b Write in any heights that are given.

Summary

There are three main methods of showing height on maps. These are spot heights, layer colouring and contours.

How do contours show height and relief?

Lines on a map that join places of the same height are called **contours**. Contours show the height of the land and what shape it is. The shape of the land is called **relief**. The difference in height between contours is chosen by the map maker. On most Ordnance Survey maps they are drawn at every 10 metres. This difference in height is called the **contour interval**. Several contours together make up a pattern. By looking carefully at these patterns you can work out how steep the slopes are and what shape the land is.

Contour lines are drawn on maps by map makers. You cannot see them on the ground. In diagram **A** the contours have been drawn on the main sketch. You will see that they make up different patterns. An important thing to remember is that:

- *the closer the contour lines are together, the steeper the slope will be.*

A

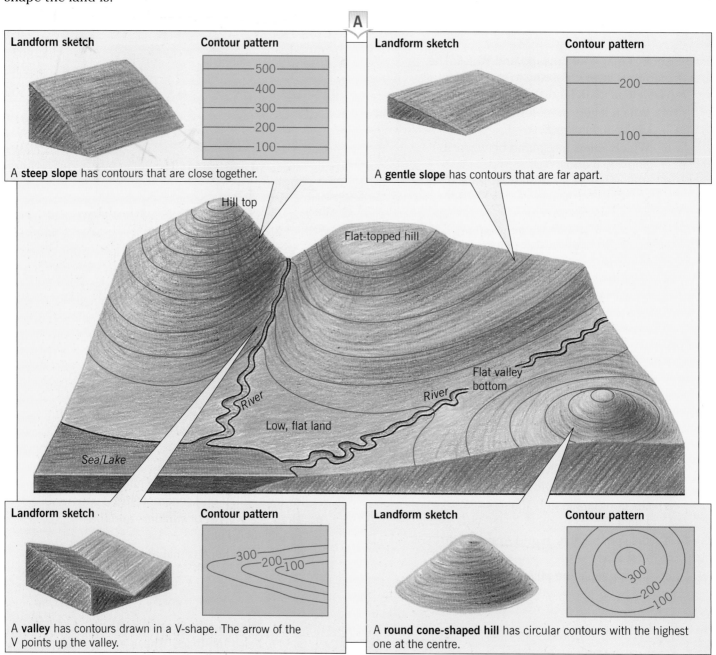

Landform sketch   **Contour pattern**

500
400
300
200
100

A **steep slope** has contours that are close together.

Landform sketch   **Contour pattern**

200

100

A **gentle slope** has contours that are far apart.

Hill top

Flat-topped hill

River

Flat valley bottom

River

Low, flat land

Sea/Lake

Landform sketch   **Contour pattern**

300 200 100

A **valley** has contours drawn in a V-shape. The arrow of the V points up the valley.

Landform sketch   **Contour pattern**

300
200
100

A **round cone-shaped hill** has circular contours with the highest one at the centre.

Activities

1 From map **B** give the heights of the following places. Choose your answers from those in the brackets.
   a The highest point is (22, 48, 52, 40, 60) metres.
   b Place **E** is (8, 42, 30, 20, 16) metres.
   c Place **B** is (30, 20, 26, 46, 34) metres.
   d Place **A** is (15, 10, 34, 6, 21) metres.
   e Place **D** is (28, 10, 12, 22, 8) metres.

2 Look at map **B** and say if the following statements are TRUE or FALSE.
   a **E** and **F** are at the same height.
   b **D** is higher than **F**.
   c **B** is higher than **E** but lower than **C**.
   d **A** is the lowest place marked with a letter.
   e **D** to **C** is steeper than **A** to **B**.

3 The photos in **C** show some landscape features.
   a Draw a simple contour pattern for each of the photos.
   b Write a description of the feature next to each of your drawings.

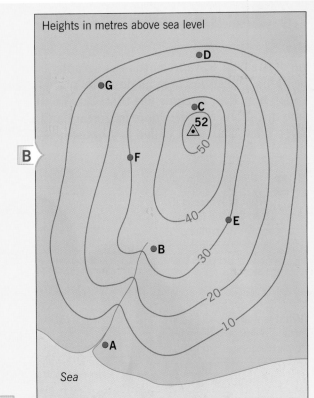

4 Look at the six letters on map **D**. Match the letters to each of the following:

   1 A gentle slope   4 A flat valley floor
   2 A steep slope   5 A valley with a stream
   3 A hill top   6 A valley without a stream.

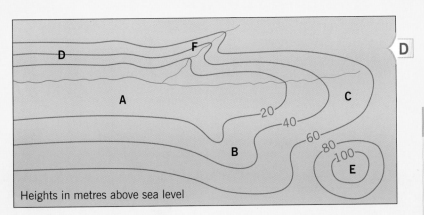

Heights in metres above sea level

Summary

Contour lines are a good way of showing height and relief on a map. Contours that are close together show steep slopes. Contours that are far apart show gentle slopes.

How can we use key questions?

One of the best ways to learn in geography is to ask questions. Questions help us describe places and explain how they became like that. Usually there are some questions that are more important and useful than others. These are called **key questions**.

Key questions may also be used to structure our writing. Writing a sentence or two in answer to each question makes it easier to describe and explain geographical features in an organised way. The questions below ask some of the things that geographers are most interested in.

A

1 What or where is it?
Describe the feature or give a location.

2 What is it like?
Include physical and human features.

3 Why is it like this?
Suggest reasons for the main features.

4 How is it changing?
Describe any changes in the landscape, buildings or land use.

5 What have been the effects of the changes?
Write about the effects on both places and people.

6 What do you think and feel about the place?
Give your view and suggest reasons for what you say.

B Scarborough, North Yorkshire

C Cliff collapse near Scarborough, Yorkshire

Activities

1 What is the difference between ordinary questions and key questions?

2 Read description **D** of Scarborough, which is shown in photo **B**. Copy and complete the description using the following words:

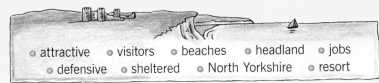

- attractive   - visitors   - beaches   - headland   - jobs
- defensive   - sheltered   - North Yorkshire   - resort

3 Look at photo **C** which shows a landslide caused by erosion near Scarborough. Describe the scene by answering the key questions in photo **A**.

4 Look back at photo **B** on page 16, which shows the effects of a storm in East Anglia.
  a Write out the six key questions from photo **A**.
  b Next to each one, write a short sentence to answer the question.

D > Scarborough is a small town in .... It is situated on the coast next to sandy .... and a steep-sided ....

The original town grew up around the castle, which had a good .... site. The .... harbour and good fishing helped the town develop in later years. Nowadays, Scarborough has become an important tourist ....

This has made it more crowded but has provided .... and new amenities which the local people can use.

Scarborough has an .... location and plenty of things for .... to do.

Summary

Key questions are important questions that help us describe and explain geographical features. They may be used to help structure our writing.

How can we describe places?

Look at photo **A** below. It shows a typical scene in the mountains of northern Italy. The labels describe the main features of the area. They include both physical and human features. There is also a conclusion which provides a neat summary of what the place is like.

Knowing how to describe a place like this is an important and useful skill in geography.

The best way to describe a place is to set out a list of geographical terms and then write a sentence or two about each one in turn. Drawing **B** shows some of the more important terms.

Not all the terms need to be used all of the time. Sometimes you may use only the ones that are important to your study.

A Dolomites, Italy

Location
It is in the mountains of northern Italy, near Cortina.

Work
There are few jobs in the area. Most people are farmers or forestry workers, or work in tourism.

Relief
There are high, jagged mountains with cliffs and steep slopes.

Population
Very few people live in the area as it is cold and mountainous.

Drainage
Many small streams flow down the steep slopes from the ice and snow.

Communications
Travel is difficult with only one narrow, winding road visible.

Vegetation
The lower slopes have thick forest and some small grassy areas.

Settlement
There are a few scattered buildings and a small village with a church.

Climate
It is cold on the high ground were there is snow on the mountains.

Conclusion
The area is very attractive to look at but it may be difficult to live and work there.

B

Physical features

 Useful terms for describing places

 Human features

 **Location**
Describes where a place is. Names of places should be used where possible.

 **Relief**
This is the height and shape of the land. Is it steep, gently sloping or flat?

Drainage
Includes all water features, such as streams, rivers, lakes and marshes.

Climate
Describes the weather conditions, such as temperature, rainfall and sunshine.

Vegetation
Describes the plant life and includes forest, grassland and farmed areas.

Population
This is about the people – how many are there and how crowded is it?

Settlement
Describes where people live. Are there towns or villages? What are they like?

 **Communications**
Describes what methods of travel are available. Is travel easy or difficult?

Work or employment
This is what people do for a living. What sort of jobs are people likely to do?

Conclusion
This gives a summary and may say what we think about the place.

Activities

1   Copy and complete the following sentences to describe the scene shown in photo **C**. Use the words given below:

- The location is Boscastle in ....
- The relief is a ......-sided, ... valley.
- The drainage is by the ...... flowing through Boscastle.
- The climate seems to be ...... , as the river is flooding.
- The vegetation is ...... and rough ....
- There is a small ...... in the river valley.
- Travel is difficult with only a narrow, winding ...... visible.
- Many people may work in the local ...... industry.
- Conclusion – write your own!

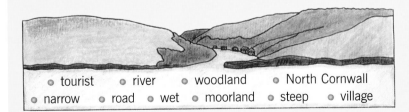

● tourist	● river	● woodland	● North Cornwall
● narrow	● road	● wet ● moorland	● steep ● village

2   Look back at photo **C** on page 47. Write a description of the scene using the terms listed as 'Human features' in drawing **B** above.

C
Flooding at Boscastle in Cornwall, UK

Summary

Describing places is an important skill in geography. It may best be done by using a list of geographical terms and writing about each one in turn.

What are key words and key sentences?

Written text in books, newspapers and magazines can be long and complicated. This can make it difficult to read and understand. To help use these sources of information in our geography studies, we must learn how to identify **key words** and **key sentences**. This can help us simplify the text and learn about the topic more easily.

These two pages look at how we can identify and use key words and key questions. This can best be done in four easy stages, as shown in drawings **A** and **B**. The example used is river flooding, a quite complicated but important topic in geography.

Key words and key sentences are the most important and most useful pieces of information in a paragraph.

A

| Read | Identify | Write | Research |

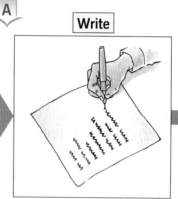

B

1 **Read** each paragraph very carefully. There is usually one important piece of information in every sentence.

2 **Identify** the key words and key sentences. In geography these may well include technical terms that are special to the subject.

3 **Write down** the key words and sentences as a list.

4 **Research** and write down the meaning of each key word and key sentence that you are not sure about. You could ask your teacher, or use a textbook or glossary for this.

Why do rivers flood?

Flooding occurs when a river <u>overflows its banks</u> and covers the surrounding area with water. This happens when the river gets <u>more water than it can hold</u>. The flooded area is called a <u>floodplain</u>.

Floods are usually caused by <u>heavy rain</u> falling over a <u>long period</u> of time. <u>Melting snow or ice</u> can also cause rivers to flood.

Most rivers have <u>always flooded</u>. This is due to <u>natural causes</u>. These include <u>steep valley slopes</u>, <u>impermeable rock or soils</u> that do not allow rain to soak into them, and <u>very wet soil</u> that no more water can soak into.

<u>Floods are more common now</u> than they used to be. Many people are blaming <u>human activity</u> for this. Two ways in which humans may increase flooding are by <u>cutting down trees (deforestation)</u> and by <u>building more towns and cities (urbanisation)</u>.

Key words and key sentences can be used to write notes. These are essential for your learning of a topic and, later, for revision. They can also be put into diagrams, which are a good way of presenting information and explaining ideas in geography.

Drawings **C** and **D** below show how such diagrams can be used to explain why rivers flood. The causes of river flooding are explained on pages 36 and 37. The diagrams will help you understand that topic.

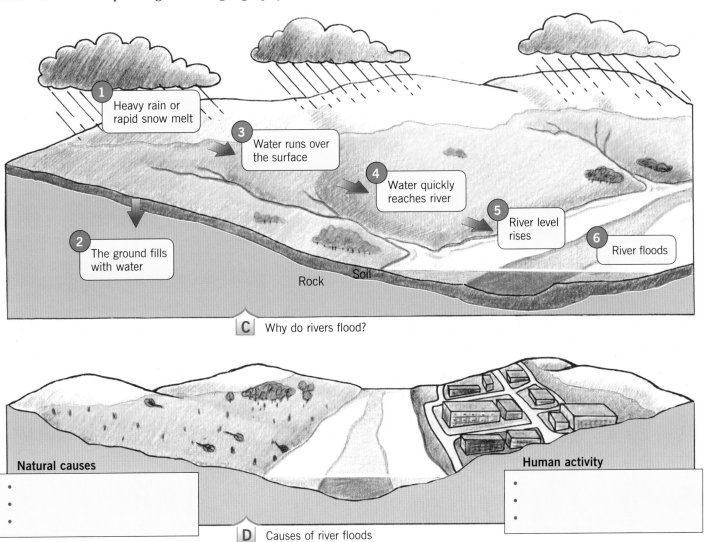

1 Heavy rain or rapid snow melt

3 Water runs over the surface

4 Water quickly reaches river

5 River level rises

6 River floods

2 The ground fills with water

Rock   Soil

C Why do rivers flood?

Natural causes

Human activity

D Causes of river floods

Activities

1 Look at photo **B**.
  a Read about why rivers flood.
  b Identify the key words and key sentences and then write them out as a list.
  c Research the meaning of the technical terms and write them out.

2 a Make a simple copy of diagram **D**.
  b Write each of the key sentences from **E** in the correct box in your diagram.

E
- Cutting down trees increases the risk of flooding.
- If soil is full of water, no more rain can soak into it.
- Impermeable rocks and soils do not allow water to soak in.
- Water flows across concrete and tarmac quickly to the river.
- Water flows quickly down steep slopes. Little soaks in.
- Gutters and drains carry water directly to the river.

Summary

The most important pieces of information in written text are known as key words or key sentences. They can be used to make notes, or added to diagrams.

How can we describe physical features on a photo?

Photos are very important in geography. They show places and features as they actually are and give us a good picture of what a place is like. Unfortunately, they often show so much detailed information that it can be difficult to identify important features. Using a **checklist** or a set of **key questions** can make the task easier.

In **physical geography** we need to describe the relief, drainage and vegetation, as shown on photo **A**.

To describe the **relief** of an area you need to look for mountains, hills and valleys and identify areas that are steep, gently sloping or flat. **Drainage features** such as lakes and rivers are easy to identify. Vegetation is a little more difficult. Grass is usually green whilst moorland and rough pasture tend to be shades of brown or yellow.

Look at the key questions below and see how you could use them to describe the physical features on photos **A** and **B**.

A Tarn Hows, Lake District, UK

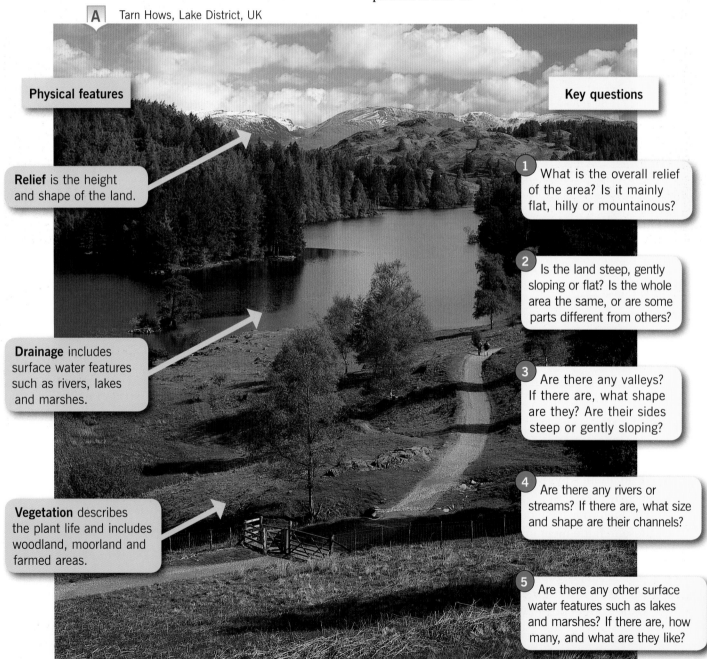

Physical features

Relief is the height and shape of the land.

Drainage includes surface water features such as rivers, lakes and marshes.

Vegetation describes the plant life and includes woodland, moorland and farmed areas.

Key questions

1 What is the overall relief of the area? Is it mainly flat, hilly or mountainous?

2 Is the land steep, gently sloping or flat? Is the whole area the same, or are some parts different from others?

3 Are there any valleys? If there are, what shape are they? Are their sides steep or gently sloping?

4 Are there any rivers or streams? If there are, what size and shape are their channels?

5 Are there any other surface water features such as lakes and marshes? If there are, how many, and what are they like?

B Maroon Bells: Aspen, Colorado, USA

Activities

1 Look at the lettered features on photo **B** above. Match the letters to each of the following relief features:

1	Steep slope	**6**	Rough grass
2	Gentle slope	**7**	Coniferous woodland
3	Mountain summit	**8**	Patch of snow
4	Valley floor	**9**	Vertical cliffs
5	Mountain lake	**10**	Scree

2 Read description **C** of the area shown on photo **B** above. Copy and complete the description using the following words:

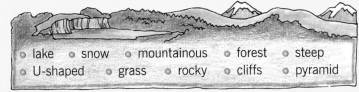

- lake ○ snow ○ mountainous ○ forest ○ steep
- U-shaped ○ grass ○ rocky ○ cliffs ○ pyramid

3 Write a paragraph to describe the physical features shown on photo **A** of Tarn Hows in the Lake District.

C

Maroon Bells: Aspen, Colorado

This part of Colorado is rugged and .... The highest peaks are .....-shaped and have patches of .... on them. There are many ...... and most of the land is very ..... The main valley is ..... with smooth, steep sides. A small ...... fills part of the valley floor.

The higher mountains are ...... and have little or no vegetation cover. The lower slopes are mainly covered in ...... with some patches of rough ....

Summary

Physical features on a photograph include relief, drainage and vegetation. They may be described using a simple checklist.

How can we use photos to study settlements?

Settlements are places where people live. Small settlements with only a few houses are called villages. Larger settlements are called towns or cities. We can find out a lot about settlements by looking at photos. They can tell us about a settlement's **site** (see page 48) and **situation**, its age, **land use pattern** and main features.

As we have seen earlier, a checklist or set of key questions can be of help when describing features on a photo. Drawing **A** is a list of key questions that you can use when looking at settlement features. Not all of the questions need to be used. It really depends on what aspects of the settlement you are most interested in.

Look at photo **B** which shows the village of Ilfracombe in Devon. The photo has been used to study the village's site and situation, as in key question 2. The labels show why a village grew up on the site.

A

1 **What is the pattern of settlement?** Are there a lot of towns and villages or only a few? Are they spread evenly throughout the area or grouped together in certain places?

2 **What are the towns and villages like?** Where are they located? What is their site and situation? Are they on high land, low land, by a river or the sea or close to resources?

3 **What is the pattern of land use?** Is the settlement crowded and congested or well planned with open space? What types of land use are there? Are there old and new areas?

4 **What communications are there?** Are there roads, railways or other forms of transport? Is the town easy to get to or are communications difficult?

B Ilfracombe, Devon, UK

Hills provide shelter from prevailing winds

Firm land provided good base for building

Nearby rock outcrops provide building materials

Fort built on steep rocky outcrop for defence

Nearby streams provided good water supply

Wood available for fuel and buildings

Good farming land nearby

Dry site well above high tide level

Sheltered harbour for fishing and trade

Sea may be used for transport

Look at photo **C** below – it shows the town of Harlow in Essex. Harlow is a new town with a population of nearly 100,000. Notice how spread out the settlement is and how modern it looks compared with Ilfracombe. What other differences can you see between Harlow and Ilfracombe?

Activities

1   Look at the lettered features on photo **C** above. Match the letters to each of these features:

   1 Main road   **4** Car park   **7** Golf course
   2 Town centre **5** Housing estate **8** Open countryside
   3 Woodland   **6** School

2   Read description **D** of the Harlow area shown on photo **C** above. Copy and complete the description using the following words:

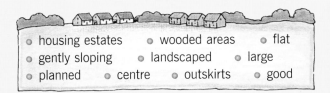

   - housing estates   - wooded areas   - flat
   - gently sloping   - landscaped   - large
   - planned   - centre   - outskirts   - good

3   Describe Ilfracombe, using key questions 2 and 3 on drawing **A**.

Description of Harlow

Harlow is a new town in Essex. It is situated on ...... or ...... land and spread out over a ...... area. There is plenty of open space and many ...... and parks.

The town looks well ...... and is neatly laid out. The ...... has many tall buildings. Surrounding the centre are modern ...... which all look very similar. An attractive, ... golf course is located on the .... The town has a ...... network of main roads.

Summary

Photos can provide us with a lot of information about settlements. They may be described using a simple checklist or a set of key questions.

113

What do aerial photos show?

Aerial photos show what the land looks like from above. They are taken from an aeroplane or helicopter and can provide accurate and detailed information about an area. Aerial photos can be vertical or oblique. **Vertical photos** are taken from directly overhead and are very similar to maps. **Oblique photos** are taken from an angle and give a better idea of the landscape's height and features – see diagram **A**.

Look at the map and aerial photo below. They both show the same area but they each show different information. The photo has a lot of detail and shows what the area really looks like. The map simplifies the detail and adds extra information like place names and what the buildings are used for.

An aerial photo shows many other features of the landscape that a map does not. It is very good, for example, at showing vegetation and crop patterns.

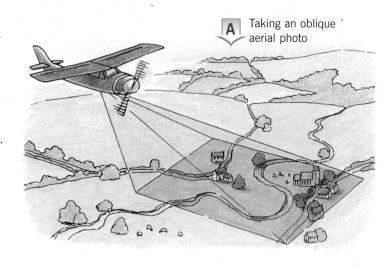

A Taking an oblique aerial photo

It can also show people and cars and suggest how busy an area was. Signs of pollution such as factory smoke and dirty rivers also show up clearly.

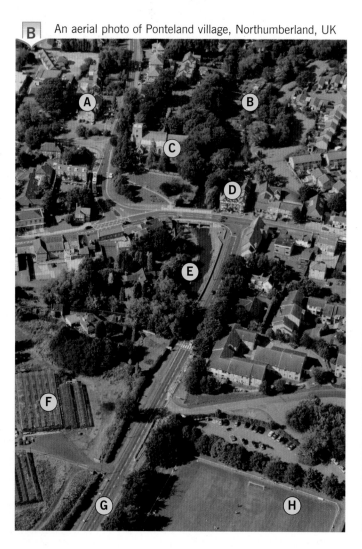

B An aerial photo of Ponteland village, Northumberland, UK

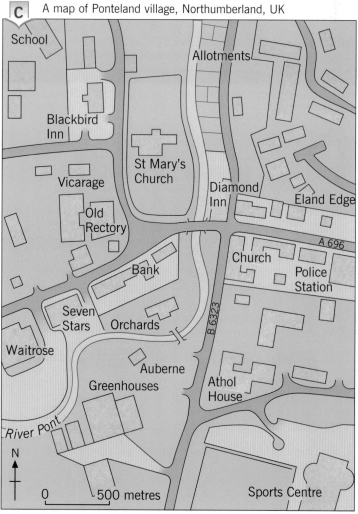

C A map of Ponteland village, Northumberland, UK

D   A vertical aerial photo of part of East London

Activities

1   Give two pieces of information that you would get:
   a   from a map but not an aerial photo
   b   from an aerial photo but not a map.

2   Look at photo **B** and map **C**.
   a   Name the features labelled A to H on the photo.
   b   Follow the course of the river on the photo. Why is it so difficult to see?
   c   Describe what you will see if you stand at the Bank in Ponteland and look north.

3   Look at photo **D** showing part of East London.
   a   Match the letters to each of the features shown in drawing **E**.
   b   From the photo, do you think the river is clean or polluted? Give reasons for your answer.

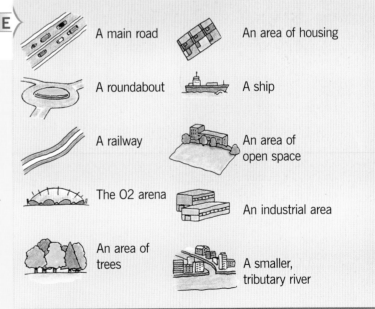

E

A main road

An area of housing

A roundabout

A ship

A railway

An area of open space

The O2 arena

An industrial area

An area of trees

A smaller, tributary river

Summary

An aerial photo shows what the landscape really looks like from above. It can provide an accurate and detailed picture of an area.

How can we use satellite photos?

Photo **B** below shows the world from space. The picture was taken from 36,000 km above the earth's surface by an orbiting satellite. The colours are natural but have been slightly changed or **enhanced**, to show features more clearly. Satellite photos like this are easy to interpret. The green areas are grassland or forest, the yellow areas desert and the white patches snow or permanent ice. All water features are coloured blue. Notice how easy it is to identify features.

Satellite photos, or **images** as they are also called, provide a vast amount of up-to-date information. They help us understand the world and make better use of its resources. Some uses of satellite information are shown in drawing **A**.

A

Some uses for satellite images in geography

- Predict weather patterns
- Help ships avoid icebergs
- Locate the best areas for fishing
- Locate new mineral and oil resources
- Identify pollution problems such as oil spills
- Assess insect damage or estimate crop yields
- Assess the impact of a natural disaster such as flooding or a volcanic eruption
- Provide information for making maps

B The world from outer space

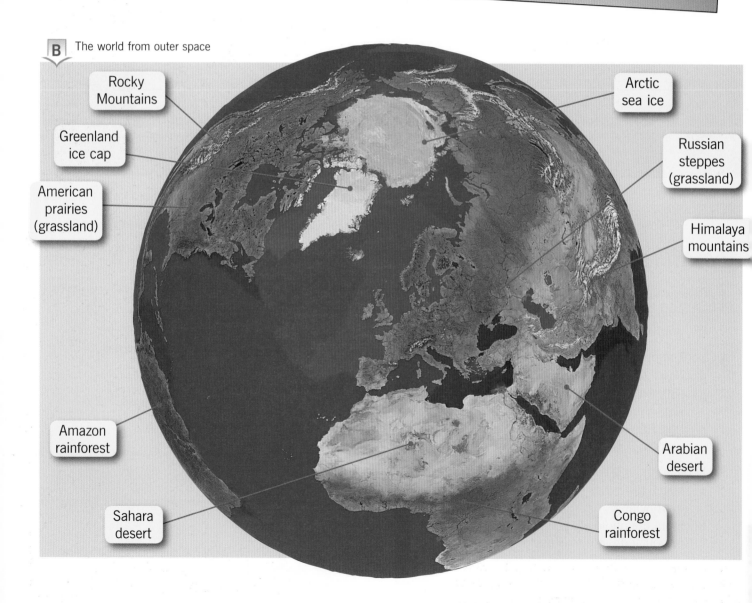

Rocky Mountains

Greenland ice cap

American prairies (grassland)

Arctic sea ice

Russian steppes (grassland)

Himalaya mountains

Amazon rainforest

Arabian desert

Sahara desert

Congo rainforest

Some satellite images use **false colours** to show information. These pictures are more difficult to understand and usually need a key. False colours make features stand out that would be hard to recognise using natural colours. For example, diseased crops, variations in sea temperatures and drought-affected areas can be shown using different colours.

Look at photo **C** which is a false colour image of San Francisco. Notice how clearly it shows the built-up areas and mountains. Can you see the Golden Gate Bridge and Bay Bridge? The earthquake zone known as the San Andreas fault runs through the mountains to the south of the photo.

C Satellite photo of San Francisco, California

Key

- Built-up area
- Areas with lush vegetation
- Drier areas with little vegetation
- Sea and San Francisco Bay

Activities

1 Many occupations can use satellite information. Match each of the following jobs with a statement from drawing **A**.

- fisherman
- farmer
- relief worker
- geologist
- weather forecaster
- surveyor
- environmentalist
- ship's captain

2 Why are false colours sometimes used on satellite images?

3 a Make a larger copy of sketch **D**.
  b Colour the sketch using the key.
  c Name the features labelled **A** to **G**. Choose from this list:

- Golden Gate Bridge
- San Andreas fault
- San Francisco Bay
- Golden Gate Park
- Montara mountains
- Airport
- San Francisco city centre

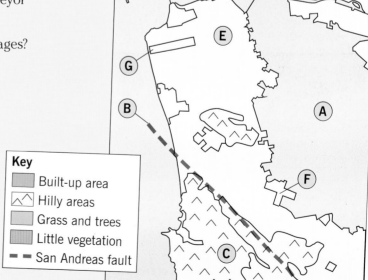

Key
- Built-up area
- Hilly areas
- Grass and trees
- Little vegetation
- San Andreas fault

Summary

Satellite photos provide an exciting way for us to look at places around the world. They can give us information that is useful in many different ways.

Glossary and Index

Accessibility How easy a place is to get to. *62*

Aerial photos Photographs taken from an aeroplane or satellite to show features on the earth's surface. *114–115*

Anticyclone A weather system with high pressure at its centre. *26–27*

Aspect The direction towards which a slope or house faces. *20*

Atmosphere The air that surrounds the earth. *6*

Beaufort scale A scale for measuring wind speed using things like smoke and trees. *18*

Birth rate The number of live births for every 1,000 of the population per year *76, 83, 85, 88–89*

By-pass A road built around a busy area to reduce traffic congestion. *68–69*

Central Business District (CBD) The middle of a town or city where most shops and offices are found. *56–57, 60, 62*

Climate The average weather conditions of a place taken over a period of years. *6, 22, 30–33, 75, 81, 106–107*

Clouds Masses of condensed water droplets suspended in the atmosphere. *18–19, 24–29*

Communications The ways by which people, goods and ideas move from one place to another. *7, 64–69, 106–107, 112*

Compass points A method of giving direction using north, south, east, west, etc. *92–93*

Condensation The process by which water vapour changes to a liquid (rain) or a solid (snow) when cooled. *24*

Congestion Overcrowding on roads causing traffic jams and delays. *62, 64–65, 67–68*

Contour A line drawn on a map joining up places with the same height above sea level. *100, 102*

Contour interval The difference in height between contours on a map. *102*

Convectional rainfall Rain that is produced when air rises after being warmed by the ground. *24–25*

Corner shop A small shop, usually in an inner city area, selling everyday things. *60, 63*

Death rate The number of deaths for every 1,000 of the population per year. *88–89*

Deforestation The cutting down of trees to clear a large area of land. *37, 108*

Depression A weather system with low pressure at its centre. *26, 28–29*

Developed country A country that has a lot of money, many services and a high standard of living. *78, 88–89*

Developing country A country that is often quite poor, has few services and a low standard of living. *72–91*

Diet The amount and type of food needed to keep a person healthy and active. *85, 87*

Direction Shown on a map by the points of the compass. *92–93*

Dispersed settlement Several isolated buildings spread out over a wide area. *50–51*

Drainage Surface water features such as rivers, lakes and marsh. *106–107, 110*

Economic activity Refers to jobs in the primary, secondary, tertiary and quaternary sectors. *7, 88, 106*

Economic geography Includes industry, jobs, trade and wealth. *7*

Embankment A raised river bank built to try to prevent flooding. *40, 42*

Environment The surroundings in which people, plants and animals live. *8–9, 15, 70–71*

Environment Agency The organisation that looks after England's rivers and coasts. *40–41*

Environmental geography The relationship between the physical and human environment. *8–9, 15*

Ethnic group A group of people with common characteristics such as race, nationality, language, religion or culture. *78, 82*

False colours On satellite images or photos, these help features to stand out that would otherwise be hard to recognise using natural colours. *117*

Flood prevention scheme A plan to try to stop flooding by either a river or the sea. *42–43*

Flooding The flow of water over land that is usually dry, e.g. a river flowing over flat land next to it, or the sea covering low-lying coastal areas. *34–45, 108–109*

Four figure grid references A group of four figures used to find a grid square on an Ordnance Survey map. *96*

Front The boundary between warm and cold air. *25, 28–29*

Frontal rainfall Rain that is produced when warm air is forced to rise over cold air in a depression. *24–25, 28–29*

Function The main purpose of a town or parts of a town, such as residential, industrial, commercial and recreational. *52, 56*

Graphs Diagrams showing information in a pictorial way, e.g. how two variables such as population growth and time may be related. *12–13*

Grid square A square on a map representing an area on the ground. *96*

Gross National Product (GNP) The wealth of a country. It is worked out by dividing the total amount of money the country earns by its total population. *88–91*

Hazard A natural event such as an earthquake, storm or flood that causes danger to people and their property. *34–45, 108–109*

Health How fit and well a person is. *88*

Height How high or low a place is above sea level measured in metres or feet. *22, 100–102*

High and low order goods Items sold in a shop are either high order, which cost a lot but are not bought very often e.g. furniture, or low order, which cost less but are bought more frequently e.g. bread. *60*

Human geography Where and how people live. *7–8, 14, 76–77, 82–83, 106–107*

Impermeable Materials that do not allow water to soak into them, for example some rock and soils. *36, 108*

Infant mortality rate The average number of deaths of children under the age of one year for every 1,000 births. *88–89*

Inner city An area of old factories and houses next to the city centre. *56–59*

Isobar A line on a map joining places with the same atmospheric pressure. *27–29*

Key question The most important question that needs to be answered to provide the most useful information. *104, 108–110*

Key words/terms The most important words in a piece of writing. In this book they are shown in a bold font. *107–109*

Land use Describes how the land in towns or the countryside is used. It includes housing, industry, farming and recreation. *52, 56–58, 112*

Landforms Natural features formed by rivers, the sea, ice, wind and volcanoes. *6*

Latitude How far north or south a place is from the Equator. *10–11, 22, 74–77*

Layer colouring Showing different heights on a map by using colours. *100*

Life expectancy The average number of years a person can expect to live. *88–91*

Linear settlement Buildings spread out in a line alongside a main road, railway or canal. *50–51*

Literacy rate The proportion of people who can read and write. *85, 88, 90–91*

Longitude How far east or west a place is from the Greenwich Meridian. *10–11, 74–77*

Map A drawing that shows part of the earth's surface as seen from directly above but at a reduced scale. *10–11, 92–103*

Meteorology The study of weather and climate. *18–19*

Microclimate The climate of a small area. *20–21*

Migration The movement of people from one place to another to live or work. *7, 82–84*

Non-renewal resources Resources that can only be used once, e.g. coal, oil. *8*

North Atlantic Drift A warm current of water that brings mild conditions to Britain. *22–23*

Nucleated settlement Buildings that are grouped closely together. *50–51*

Oblique air photographs Photographs taken from the air at an angle. *114*

Ordnance Survey The official government organisation responsible for producing maps in the UK. *51, 92–103, inside back cover*

Pastoralists Farmers who look after herds or flocks of animals. *86*

Pattern How features like settlement, shops and forest are spread out over an area of land. *50–51, 56, 102, 112*

Physical geography Natural features and events such as landforms and weather. *6, 8, 14, 74–75, 80–81, 106–107, 110*

Plan A detailed map of an area. *92, 114*

Pollution Harmful substances produced by people and machines which spoil the land, water and air. *8–9, 63–65, 67, 84*

Glossary and Index

'Poor' South Developing countries with little wealth and a low standard of living located mainly in Africa, parts of Asia and South and Central America. *89–91*

Population People who live in an area. *7, 76–77, 82–83, 88, 106–107*

Population growth The increase in the number of people living in an area. *88*

Precipitation Water in any form that falls to earth. It includes rain, snow, sleet and hail. *18, 23–25, 28–29, 31, 36, 75, 81–82*

Pressure Atmospheric pressure is the weight of air pressing down onto the earth's surface. *26–29*

Public transport Transport provided for the public and available to everyone. *66–67*

Quality of life A measure of how content people are with their lives and the environment in which they live. *7, 54, 76, 90*

Relief The shape of the land surface and its height above sea level. *78, 80, 100–102, 106–107, 110*

Relief rainfall Rain that is produced when warm, moist air is forced to rise over hills and mountains. *24*

Renewable resources Resources that can be used over and over again, e.g. energy from the wind. *8*

Resources Things that are useful to people. They may be natural like coal and oil or of other value like money and skilled workers. *8, 77*

Ribbon development Settlements with a long narrow linear shape beside a main road, railway or canal. *50–51*

'Rich' North Developed countries with considerable wealth and a high standard of living mainly found in North America, western Europe, Australasia and eastern Asia. *89*

Rift valley A deep, steep-sided valley formed by the sinking of the land between two faults or cracks in the earth's surface. *75, 80*

Rural-to-urban migration The movement of people from the countryside to towns. *83*

Satellite image A photo taken from a satellite orbiting in space and sent back to earth. The images can show either true, false or enhanced colours. *26, 28, 74–75, 116–117*

Scale The link between the distance on a map and its real distance on the ground. *94*

Settlement A place where people live and work. *7, 46–59, 84–85, 106–107, 112–113*

Sewage Waste material from homes and industry. *9, 84–85*

Shanty settlement A collection of shacks and poor-quality housing which often lacks electricity, a water supply and sewage disposal. *84–85*

Shopping malls Places to shop that are under cover, protected from the weather and are traffic free. *62–63*

Site The place where a settlement or a factory is located. *48–49, 112*

Six figure grid reference A group of six figures used to give an exact position on a map. *98*

Spot height A point on a map giving its height above sea level in metres. *100*

Standard of living How well-off a person or country is. *76, 78, 88–90*

Suburbanised village A village with many new buildings added to it. *52*

Suburbs A zone of housing around the edge of a city. *56–57*

Symbol A simple drawing or sign used to give information and to save space on a map. *19, 26, 28–29, 99, inside back cover*

Temperature A measure of how warm or cold it is. *18, 20–22, 31, 75, 81–82*

Trade The sale and movement of goods between countries. *7, 88–91*

Transport Ways of moving people and goods from one place to another. *7, 62–69*

Urban An area of land which is mainly covered in buildings. *54–57, 84–85*

Urban model A simple pattern to show the usual land use in a city. *56–57*

Urbanisation The growing proportion of people living in towns and cities. *37, 54–55, 70–71, 108*

Vertical air photograph A photo taken from directly overhead and resembling a map. *114–115*

Visibility The distance that can be seen. *19*

Volcanic activity Lava and ash ejected from the summit of a mountain. *75, 80*

Weather The day-to-day condition of the atmosphere. It includes temperature, precipitation, pressure and wind. *6, 16–33*

Wind speed and direction The strength of the wind and the direction from which it blows. *18–22, 27–29, 32*